AI Agents for Secure and Software-Defined Networking

Harnessing AI and SDN to Revolutionize Modern Work Environments

Het Mehta

Apress®

AI Agents for Secure and Software-Defined Networking: Harnessing AI and SDN to Revolutionize Modern Work Environments

Het Mehta
Sunnyvale, CA, USA

ISBN-13 (pbk): 979-8-8688-2357-2

ISBN-13 (electronic): 979-8-8688-2358-9

https://doi.org/10.1007/979-8-8688-2358-9

Managing Director, Apress Media LLC: Welmoed Spahr
Acquisitions Editor: Celestin Suresh John
Editorial Assistant: Gryffin Winkler

Cover designed by eStudioCalamar

Distributed to the book trade worldwide by Springer Science+Business Media New York, 1 New York Plaza, New York, NY 10004. Phone 1-800-SPRINGER, fax (201) 348-4505, e-mail orders-ny@springer-sbm.com, or visit www.springeronline.com. Apress Media, LLC is a Delaware LLC and the sole member (owner) is Springer Science + Business Media Finance Inc (SSBM Finance Inc). SSBM Finance Inc is a **Delaware** corporation.

For information on translations, please e-mail booktranslations@springernature.com; for reprint, paperback, or audio rights, please e-mail bookpermissions@springernature.com.

Apress titles may be purchased in bulk for academic, corporate, or promotional use. eBook versions and licenses are also available for most titles. For more information, reference our Print and eBook Bulk Sales web page at http://www.apress.com/bulk-sales.

Any source code or other supplementary material referenced by the author in this book is available to readers on GitHub. For more detailed information, please visit https://www.apress.com/gp/services/source-code.

If disposing of this product, please recycle the paper

Table of Contents

Chapter 8: Reinforcement Learning for Autonomous Traffic Engineering 71

Chapter 9: Ethical AI Agents for Responsible Network Management 83

Chapter 10: AI Agents for Network Threat Intelligence and Response 89

About the Author

Het Mehta is a Software Developer at Cisco, revolutionizing network infrastructure through SD-WAN, security automation, and artificial intelligence. His work bridges the gap between traditional networking and intelligent, software-defined systems that adapt to the demands of modern enterprises.

With deep expertise spanning Software-Defined Networking, cloud integration, and kernel-level development, Het has designed secure SD-WAN solutions that optimize performance while reducing operational complexity. His unique background in embedded systems gives him a rare perspective on how hardware and software intersect to create resilient, high-performance networks.

Now at the forefront of AI-driven networking, Het is pioneering intelligent systems that enable proactive threat detection, autonomous configuration, and dynamic resource management. His vision: networks that don't just respond to change—they anticipate it.

Through this book, Het shares his insights on how AI agents are transforming secure and software-defined networking, offering practical frameworks for building the autonomous networks of tomorrow. Whether you're a network engineer, security professional, or technology leader, Het's expertise will guide you through the next evolution of networking infrastructure.

About the Technical Reviewer

Nikita Kothari is a Senior Member of Technical Staff at Salesforce, where she builds AI-driven enterprise solutions that integrate Large Language Models (LLMs), agentic AI, and intelligent automation to enhance personalization, access, and system trust. With over a decade of experience spanning Salesforce, LinkedIn, and Amazon, she has led transformative projects in AI messaging, recommendation systems, and scalable data synchronization frameworks. Nikita holds a Master of Science in Computer Science from the University of Texas at Dallas and is passionate about advancing responsible and practical AI applications for modern enterprise ecosystems.

Introduction

This book explores how Artificial Intelligence (AI) and Software-Defined Networking (SDN) transform the way modern networks are designed, secured, and operated. In an era shaped by cloud computing, IoT, 5G, and edge computing, traditional network management is no longer sufficient—this book reveals how AI agents bring autonomy, intelligence, and adaptability to meet these challenges.

Beginning with foundational concepts in AI and SDN, the book guides readers toward advanced architectures and real-world applications. It addresses critical needs such as scalable, self-healing networks and proactive cybersecurity, demonstrating how AI techniques—including reinforcement learning, graph neural networks, and explainable AI—enable intent-based networking, cognitive healing, federated learning, and intelligent automation.

Each chapter combines conceptual overviews with detailed discussions, case studies, and actionable insights, making it accessible to students, researchers, engineers, and decision-makers alike. Beyond technical depth, it addresses ethics, governance, energy efficiency, and disaster recovery, unifying AI and networking into a single, practical framework. Curated resources—including books, blogs, courses, and glossaries—support ongoing learning beyond the text.

This book serves as a comprehensive road map for designing intelligent, secure, and adaptive network ecosystems—essential for those leading the next generation of decentralized, resilient, and AI-driven digital infrastructure.

What You Will Learn

- Understand core AI-agent architectures and their integration with SDN for scalable, adaptive network environments.

- Apply ML, deep learning, reinforcement learning, and graph neural networks to optimize, automate, and secure networks.

- Develop AI-enabled networks using real-world case studies from telecom, smart cities, and enterprise IT.

- Explore emerging trends like federated learning, edge AI, programmable optical networks, and AI-driven disaster recovery.

Preface

In today's rapidly evolving digital landscape, the fusion of Artificial Intelligence (AI) and Software-Defined Networking (SDN) is ushering in a new era of intelligent, adaptive, and secure network architectures. This convergence is not only transforming how networks are designed, managed, and optimized but is also fundamentally reshaping the way organizations operate and engage with their most critical digital assets. The book you hold, *AI Agents for Secure and Software-Defined Networking: Harnessing AI and SDN to Revolutionize Modern Work Environments*, presents an in-depth, holistic exploration of this transformative journey.

As the volume, velocity, and variety of data continue to grow exponentially—fueled by cloud computing, Internet of Things (IoT), 5G, edge computing, and increasingly sophisticated cyber threats—traditional network management paradigms struggle to keep pace. Static configurations, manual interventions, and siloed security measures are no longer sufficient to meet the demands of modern, distributed, and dynamic network environments. Simultaneously, AI technologies have matured to a point where they can imbue networks with unprecedented levels of autonomy, intelligence, and resilience.

This book charts a comprehensive course through the multifaceted world of AI-driven SDN, beginning with foundational principles and progressing through advanced architectures, practical applications, and future trends. It provides readers with the knowledge, frameworks, and actionable insights necessary to harness AI agents for decentralized network management, intent-based networking, federated learning, cross-layer orchestration, digital twins, cognitive healing, dynamic policy management, and much more.

What You Will Discover

- **Foundational Concepts and Architectures:** Understand the core building blocks of AI-agent architectures, their types, and how decentralization enhances scalability, resilience, and adaptability in complex network ecosystems.

- **Advanced AI Techniques:** Explore machine learning, deep learning, reinforcement learning, graph neural networks, natural language processing, and explainable AI, all tailored to optimize network performance, security, and automation.

- **Real-World Applications and Case Studies:** Gain practical insights from diverse industries and scenarios—from smart cities and telecom operators to enterprise SD-WAN deployments—illustrating how AI agents deliver measurable improvements in efficiency, security, and user experience.

- **Ethical and Governance Considerations:** Delve into the critical ethical frameworks, data governance models, and regulatory compliance strategies that ensure AI deployment in networking is responsible, transparent, and trustworthy.

- **Future Trends and Emerging Technologies:** Stay ahead with discussions on federated learning, edge AI, AI-powered energy efficiency, multi-domain collaborative security, programmable optical networks, and AI-driven disaster recovery.

- **Comprehensive Resources:** Benefit from curated learning materials, recommended books, influential blogs, online courses, research papers, and a detailed glossary of essential terminologies to support your ongoing mastery.

Who Should Read This Book

This book is crafted to serve a broad and diverse audience, including but not limited to

- **Network Architects and Engineers:** Professionals designing and managing modern networks who seek to integrate AI-driven automation and intelligent control to overcome scalability and complexity challenges

- **AI Researchers and Data Scientists:** Specialists interested in applying advanced AI methodologies to networking problems, from anomaly detection to reinforcement learning for traffic engineering

- **IT and Business Leaders:** Executives and decision-makers aiming to leverage AI and SDN technologies to drive operational agility, reduce costs, enhance security, and foster innovation within their organizations

- **Security Professionals:** Experts focused on safeguarding networks who want to understand AI's role in proactive threat detection, incident response, and collaborative security across multi-domain environments

- **Academics and Students:** Scholars and learners exploring the intersection of AI, networking, and cybersecurity, looking for a comprehensive and up-to-date resource

- **Technology Implementers and Vendors:** Practitioners developing or deploying AI-enabled SDN solutions, seeking best practices, architectural guidance, and real-world insights

Why This Book Matters

The integration of AI and SDN is not merely a technological upgrade; it represents a fundamental paradigm shift in how networks operate and support business and societal needs. AI agents empower networks to move beyond static configurations and reactive management toward systems that are self-aware, self-learning, and self-optimizing.

This transformation enables

- **Scalability at Unprecedented Levels:** Distributed AI agents that collaborate across different locations and manage vast, heterogeneous networks efficiently, overcoming bottlenecks inherent in centralized control.

- **Enhanced Security Posture:** AI-driven anomaly detection, threat intelligence sharing, and automated mitigation reduce risks in complex, distributed environments.

- **Operational Efficiency and Cost Savings:** Automated provisioning, dynamic policy management, and energy-aware networking optimize resource usage and reduce manual workload.

- **Improved User Experience:** Personalized network management and intent-based networking ensure that services dynamically align with user needs and business goals.

- **Resilience and Business Continuity:** Cognitive healing and AI-powered disaster recovery prepare networks to anticipate, withstand, and rapidly recover from disruptions.

By embracing these capabilities, organizations can build the next generation of workplaces—agile, secure, intelligent, and responsive to the evolving digital ecosystem.

How This Book Is Structured

The book is organized into thematic chapters, each focusing on a critical aspect of AI-driven SDN and networking:

- **Chapters 1–6:** Lay the foundation by introducing AI-agent architectures, intent-based networking, federated learning, cross-layer orchestration, digital twins, and cognitive healing.

- **Chapters 7–14:** Dive into dynamic policy management, reinforcement learning for traffic engineering, ethical AI deployment, threat intelligence, resource allocation, edge-centric management, network function chaining, and zero-touch provisioning.

- **Chapters 15–21:** Explore graph neural networks, programmable data plane security, personalized network management, energy efficiency, collaborative security in multi-domain SDNs, programmable optical networks, and predictive analytics for scalability.

- **Chapters 22–25:** Highlight the democratization of AI-driven SDN management through open source, SD-WAN optimization and security, and AI-enabled disaster recovery.

- **Chapter 26:** Concludes with additional resources for further learning and a comprehensive glossary of key terms.

Each chapter begins with an introduction, develops major concepts with clear headings, includes real-world examples and case studies, and ends with conclusions and forward-looking insights. This structure ensures a logical flow and makes the book accessible to readers with varying levels of expertise.

How to Use This Book

Whether you are reading cover-to-cover, referencing specific chapters, or using it as a guide for implementation, this book is designed to be a practical and authoritative resource. The inclusion of case studies, best practices, and curated resources supports both theoretical understanding and hands-on application.

For educators and trainers, the book offers a rich curriculum foundation. For practitioners, it provides actionable frameworks and insights. For leaders, it delivers strategic perspectives to guide AI and SDN adoption.

Final Thoughts

As networks become the nervous system of our digital world, the integration of AI and SDN will be pivotal in shaping resilient, intelligent, and user-centric infrastructures. This book aims to equip you with the knowledge and tools to navigate and lead in this exciting frontier.

Together, we stand at the cusp of a new era—where networks are not just connected but truly intelligent, adaptive, and secure. I invite you to join this journey and help shape the future of workplaces and digital ecosystems empowered by AI-driven Software-Defined Networking.

Welcome to the future. Let's explore it together.

CHAPTER 1

AI-Agent Architectures for Decentralized Network Management

The modern digital world demands networks that are not only vast but also highly dynamic and complex. With the proliferation of IoT devices, cloud services, and mobile connectivity, network infrastructures have grown exponentially in size and intricacy. Traditional network management approaches, which rely heavily on centralized control and manual configuration, are increasingly inadequate for handling this complexity efficiently.

Centralized systems often become bottlenecks, unable to scale effectively or respond swiftly to localized issues. Moreover, they present single points of failure, risking network-wide outages if the central controller is compromised or overwhelmed. To overcome these challenges, decentralized network management architectures empowered by AI agents have emerged as a promising solution.

> *The future of networking lies in intelligent, autonomous agents that can manage complexity without centralized control—a true revolution in how networks think and adapt.*

AI agents are intelligent software entities capable of sensing their environment, making autonomous decisions, and acting to optimize network performance, security, and reliability. When deployed across a distributed network, these agents collaborate to manage resources, detect anomalies, and adapt to changing conditions without requiring constant human oversight.

This chapter delves into the core principles of AI-agent architectures designed for decentralized network management. We will explore the components that constitute these agents, the types of architectures in use, and the tangible benefits decentralization brings to modern networks. Through real-world case studies, the chapter illustrates how these concepts manifest in practical deployments. Finally, it surveys future trends that are shaping the evolution of AI-driven decentralized network management.

Overview of AI-Agent Architectures

To understand how decentralized AI-agent architectures work, it is important to first define what AI agents are and how they function within a network. This section also introduces the key components of AI-agent architectures and the various types of agents deployed in decentralized systems.

Defining AI Agents in Networking

An AI agent in the context of networking is a software entity that operates autonomously to perform tasks such as monitoring network traffic, detecting faults, optimizing routing paths, and managing security policies. Unlike traditional scripts or static software, AI agents leverage machine learning, reasoning, and decision-making algorithms to adapt their behavior based on observed network states.

The autonomy of AI agents allows them to function with minimal human intervention, making real-time decisions that enhance network efficiency and resilience. They can be programmed to pursue specific goals, such as minimizing latency, maximizing throughput, or ensuring compliance with security policies.

Core Components of AI-Agent Architectures

To function effectively, AI agents incorporate several interrelated components:

- **Perception Module:** This component collects data from network devices, sensors, and logs. It includes mechanisms for real-time monitoring of bandwidth usage, latency, packet loss, and security events. The quality and timeliness of data gathered here directly influence the agent's decision-making capabilities.

- **Decision Engine:** The heart of the agent, this module processes the perceived data using AI algorithms. Techniques such as supervised learning, reinforcement learning, and heuristic methods enable the agent to predict network behavior, classify events, and select optimal actions.

- **Action Module:** Once a decision is made, this component implements the necessary changes on the network. Actions might include adjusting routing tables, reallocating bandwidth, triggering alerts, or isolating suspicious nodes.

- **Communication Interface:** AI agents rarely operate in isolation. This module enables communication with other agents and network elements, facilitating collaboration, information sharing, and consensus-building for coordinated management.

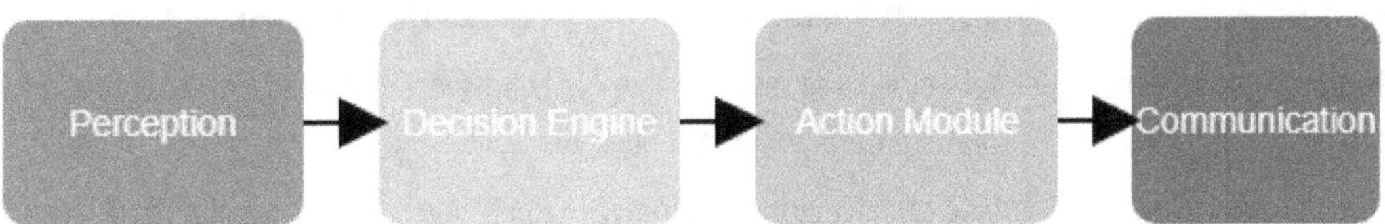

Figure 1-1. *AI-Agent Architecture Core Components*

Types of AI-Agent Architectures

AI agents vary in complexity and operational style. The following types are prevalent in decentralized network management:

- **Reactive Agents:** These agents respond directly to stimuli without maintaining an internal model of the environment. They are fast and simple, suitable for immediate event handling like packet loss detection or link failure response.

- **Deliberative Agents:** Unlike reactive agents, deliberative agents maintain an internal state or model of the network. They plan their actions based on predictions and long-term goals, such as optimizing routing paths over time or forecasting traffic congestion.

- **Hybrid Agents:** Combining reactive and deliberative traits, hybrid agents can respond quickly to urgent events while also engaging in strategic planning. This dual capability is beneficial in scenarios like real-time security threat mitigation.

- **Collaborative Agents:** Unlike the agents discussed so far, these agents work in concert with peers, sharing information and coordinating actions to achieve global objectives. Examples include distributed load balancing and coordinated intrusion detection across multiple network segments.

Table 1-1. *Types of AI-Agent Architectures and Their Use Cases*

Architecture Type	Description	Example Use Case
Reactive Agents	Respond to immediate stimuli without internal state.	Simple and real-time anomaly detection and response
Deliberative Agents	Maintain models of the environment for planning actions.	Long-term traffic pattern optimization
Hybrid Agents	Combine reactive and deliberative capabilities.	Real-time security threat mitigation with feedback
Collaborative Agents	Coordinate with peers to achieve global objectives.	Distributed load balancing coordination

Decentralization Benefits in Networking

After understanding the components and types of AI agents, it is essential to explore why decentralization is critical for modern network management. This section highlights the challenges of centralized systems and the advantages of decentralized AI-agent architectures.

Challenges of Centralized Network Management

Centralized management systems have been the backbone of traditional network control. However, they face significant challenges in modern environments:

- **Scalability Bottlenecks:** As networks expand, centralized controllers must process increasing volumes of data and commands, leading to performance degradation.

- **Latency Issues:** Centralized decision-making can introduce delays, especially when controllers are geographically distant from the data sources or affected nodes.

- **Single Point of Failure:** Centralized controllers represent critical points whose failure can disrupt the entire network, causing outages or degraded service.

- **Limited Flexibility:** Centralized policies may not quickly adapt to local conditions or emerging threats, reducing responsiveness.

Advantages of Decentralized AI-Agent Architectures

Decentralized AI-agent architectures address these limitations by distributing intelligence throughout the network:

- **Improved Scalability:** By delegating control to multiple autonomous agents, the network can scale organically without overwhelming any single component.

- **Lower Latency:** Agents make decisions locally, enabling faster reactions to network changes, such as rerouting traffic around congested links.

- **Enhanced Resilience:** The network becomes more fault-tolerant, as the failure of one agent affects only its local domain rather than the entire system.

- **Context-Aware Adaptation:** Agents tailor their decisions based on localized data, leading to more precise and effective management.

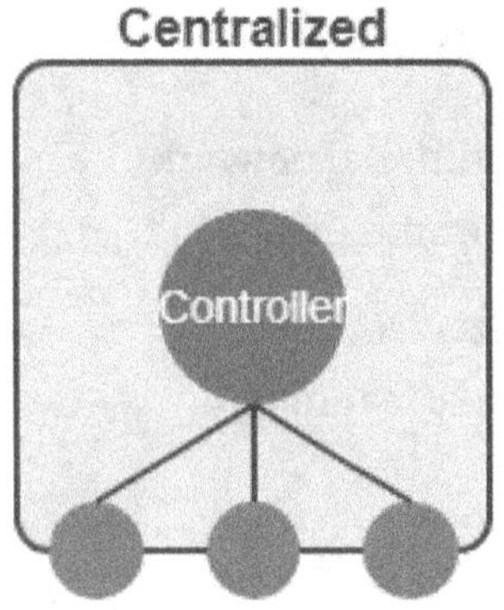

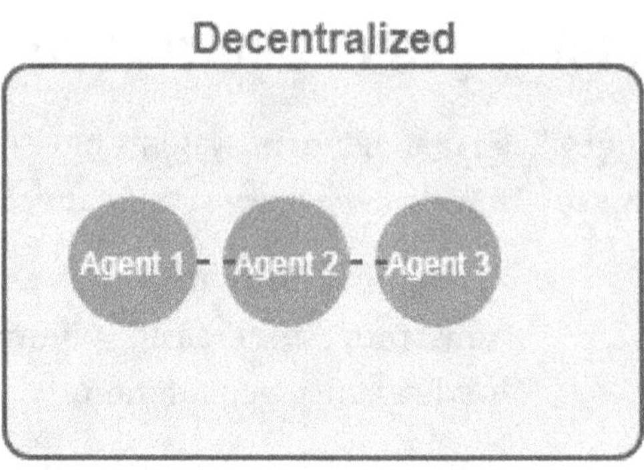

Figure 1-2. *Centralized vs. Decentralized Network Management Architectures*

Table 1-2. *Comparison Between Centralized and Decentralized Network Management*

Feature	Centralized Management	Decentralized AI-Agent Management
Scalability	Limited by central controller	High, distributed across agents
Latency	Higher due to centralized control	Low, decisions made locally
Fault Tolerance	Single point of failure risk	High, agents operate independently
Flexibility	Rigid, slower policy updates	Adaptive, dynamic policy enforcement
Complexity Handling	Challenging with large networks	Better suited for complex, dynamic networks

Case Studies in Decentralized Management

To illustrate the practical impact of decentralized AI-agent architectures, this section explores three real-world case studies.

Case Study 1: Autonomous Traffic Routing in Smart Cities

Smart cities face the challenge of managing traffic congestion and ensuring smooth transit for emergency services. Traditional traffic management systems rely on centralized control centers, which often suffer from latency and scalability issues.

In one metropolitan deployment, AI agents are embedded at local traffic signal controllers. These agents monitor vehicle density and traffic flow in real time using sensors and cameras. By communicating with neighboring agents, they dynamically adjust signal timings to optimize traffic movement and prioritize emergency vehicles.

This decentralized approach has led to a measurable reduction in average commute times and improved emergency response efficiency, demonstrating the effectiveness of AI-agent architectures in complex urban environments.

Case Study 2: Multi-agent Security Monitoring in Enterprise Networks

Enterprises with distributed branch offices face challenges in timely detection and response to security threats. Centralized security operations centers can be overwhelmed by the volume of data and may lack visibility into local anomalies.

A multinational corporation implemented a network of collaborative AI agents across its offices. Each agent monitors local network traffic for suspicious patterns and shares threat intelligence with peers. This peer-to-peer collaboration enables rapid identification of coordinated attacks and containment at the source.

The system has improved incident response times by over 40% and reduced false positives, showcasing the power of decentralized AI in enhancing cybersecurity.

Case Study 3: Distributed Resource Management in Data Centers

Data centers require efficient resource allocation to balance performance and energy consumption. Centralized management can lead to suboptimal decisions due to delayed or aggregated data.

AI agents deployed on individual servers and racks continuously monitor workload and power usage. Through local decision-making and inter-agent communication, they redistribute workloads dynamically to optimize energy efficiency without compromising service quality. This decentralized approach has resulted in up to 20% energy savings and improved server utilization rates.

Future Trends in AI Architectures

As AI architectures evolve, one key area of advancement is their deployment beyond centralized cloud systems.

Integration with Edge and Fog Computing

Edge and fog computing paradigms bring computation and intelligence closer to data sources, reducing latency and bandwidth usage. AI agents integrated into edge devices will enable real-time, context-aware network management, particularly for IoT and 5G applications.

Advances in Multi-agent Collaboration

Emerging research focuses on enhancing collaboration among AI agents through improved communication protocols and consensus algorithms. Techniques such as blockchain-based trust mechanisms and federated learning will enable secure, scalable cooperation.

Explainable AI in Network Management

As AI agents assume greater control, the need for transparency grows. Explainable AI (XAI) techniques will allow operators to understand agent decisions, facilitating trust, debugging, and compliance with regulatory requirements.

Security and Privacy Enhancements

Future AI-agent architectures will embed robust security measures to protect against adversarial attacks and unauthorized access. Privacy-preserving AI models will ensure sensitive data remains confidential while enabling effective learning and decision-making.

Conclusion

The shift from centralized to decentralized network management powered by AI-agent architectures marks a significant evolution in how networks are controlled and optimized. By distributing intelligence and autonomy across multiple agents, networks gain scalability, resilience, and adaptability necessary for modern demands.

This chapter has provided a comprehensive overview of AI-agent architectures, highlighting their components, types, and benefits. Through illustrative case studies, the practical advantages of decentralization are evident. Looking forward, the integration with emerging technologies such as edge computing and explainable AI promises to further enhance decentralized network management.

The next chapter will build on this foundation by examining how adaptive AI agents enable intent-based networking, allowing networks to interpret and fulfill user intentions autonomously.

CHAPTER 2

Adaptive AI Agents for Intent-Based Networking (IBN)

The complexity of modern networks has given rise to the need for intent-based networking (IBN), a paradigm that shifts the focus from manual configurations to high-level intent. IBN enables network operators to define what they want the network to achieve, and the system automatically translates these intents into actionable configurations. Central to this transformation are adaptive AI agents, which interpret user intent, dynamically adjust network behavior, and ensure continuous alignment with desired outcomes.

The next evolution of networking lies in understanding and delivering user intent seamlessly, powered by adaptive AI agents.

This chapter explores the foundational concepts of IBN, the role of AI in enabling this paradigm, and the challenges associated with its implementation. We also examine how AI-driven systems recognize and adapt to user intent, ensuring efficient and reliable network operations.

Understanding Intent-Based Networking

Intent-based networking represents a significant departure from traditional network management approaches. Instead of manually configuring devices, operators define their goals in plain language or high-level policies. For example, an intent might be: "Ensure maximum bandwidth for video conferencing during business hours." The network then interprets this intent and configures itself to meet the specified requirements.

11

© Het Mehta 2026
H. Mehta, *AI Agents for Secure and Software-Defined Networking,*
https://doi.org/10.1007/979-8-8688-2358-9_2

Key Components of IBN

The key components of IBN include

- **Intent Definition:** Operators express their desired outcomes in a clear and abstract manner. This involves translating business objectives into network policies without delving into technical specifics.

- **Intent Translation:** The system converts high-level intents into specific network configurations and policies. This involves parsing the intent, understanding the context, and generating the necessary commands for network devices.

- **Continuous Assurance:** AI agents monitor the network to ensure that the system continuously meets the defined intent, adapting to changes in real time. This involves ongoing validation and adjustment to network configurations to maintain alignment with user intent.

The IBN workflow is a cyclical process that begins with intent input and ends with continuous assurance, as shown below. This enhanced diagram illustrates the interactions between various stages more clearly.

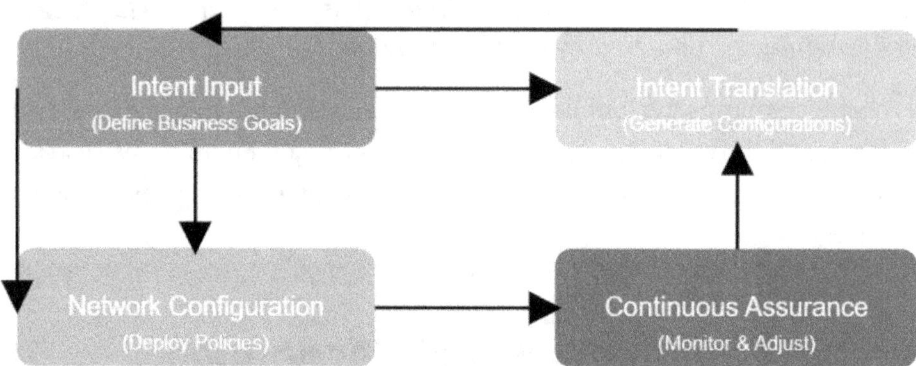

Figure 2-1. *The Enhanced Intent-Based Networking Workflow*

By automating these processes, IBN reduces the complexity of managing large-scale networks while enhancing agility and reliability.

Role of AI in IBN

The success of intent-based networking hinges on the capabilities of AI. Adaptive AI agents play a central role in interpreting intent, implementing configurations, and ensuring continuous alignment with the desired outcomes.

AI Capabilities in IBN

AI contributes to IBN in the following ways:

- **Intent Interpretation:** AI systems use natural language processing (NLP) to understand user-defined intents, even when expressed in ambiguous or incomplete terms. This capability is crucial for accurately capturing the essence of user goals and translating them into actionable tasks.

- **Policy Generation:** Machine learning algorithms translate high-level intents into actionable network policies tailored to the current network state. This involves leveraging historical data and predictive models to optimize configurations.

- **Real-Time Adaptation:** AI agents continuously monitor the network, detecting deviations from the intended state and making adjustments as needed. This ensures that the network remains aligned with user intentions, even as conditions change.

- **Predictive Analytics:** By analyzing historical data, AI predicts potential issues and proactively adjusts configurations to prevent disruptions. This preemptive approach enhances network reliability and performance.

The integration of AI into IBN enables networks to operate with minimal human intervention while maintaining high levels of performance and reliability.

Implementation Challenges and Solutions

While the benefits of IBN are compelling, implementing this paradigm poses several challenges. These include technical, operational, and organizational barriers. However, with the right strategies, these obstacles can be overcome.

Challenge 1: Intent Ambiguity

Users often express intents in vague or incomplete terms, making it difficult for the system to interpret and translate them accurately. For example, "optimize performance" can mean different things depending on the context.

Solution: AI systems equipped with advanced NLP and context-awareness can clarify ambiguous intents by asking follow-up questions or analyzing historical patterns. Continuous learning enables the system to improve its understanding over time.

Challenge 2: Legacy Infrastructure

Many organizations operate hybrid environments that include legacy systems not designed for IBN. Integrating these systems into an intent-based framework can be complex.

Solution: Deploying AI agents as middleware allows legacy systems to communicate with modern IBN platforms. These agents act as a protocol translation layer, converting modern SDN/IBN commands into legacy protocols (SNMP, CLI) that older devices understand while also translating device responses back into formats compatible with IBN systems.

Challenge 3: Scalability

As networks grow, the volume of intents and configurations increases, potentially overwhelming the system.

Solution: Distributed AI-agent architectures ensure scalability by delegating control to multiple autonomous agents. These agents collaborate to manage local segments of the network, reducing the burden on central systems.

Challenge 4: Security Concerns

Automated intent execution introduces risks, as malicious actors could exploit vulnerabilities to manipulate network configurations.

Solution: AI-driven security mechanisms, such as anomaly detection and policy validation, safeguard the system against unauthorized changes. Blockchain-based intent logging ensures transparency and accountability. This ensures immutability and audit trails, while complementary AI-driven anomaly detection provides real-time threat prevention.

AI-Driven User Intent Recognition

Understanding user intent lies at the heart of IBN. AI systems must not only interpret what users explicitly state but also infer their implicit needs and preferences. This requires a combination of natural language processing, contextual analysis, and machine learning.

How AI Recognizes User Intent

- **Natural Language Processing (NLP):** AI systems process user input, identifying key phrases and extracting actionable information. This involves parsing complex language structures and understanding nuances.

- **Contextual Analysis:** The system considers the current network state, user roles, and historical patterns to interpret intent accurately. By evaluating the broader context, AI can discern the underlying goals behind user requests.

- **Machine Learning Models:** These models predict user preferences and anticipate future needs based on past behavior. By continuously learning from interactions, AI systems refine their ability to recognize and fulfill user intent.

AI-Driven Intent Recognition Process
The diagram below illustrates the stages involved in recognizing user intent.
Enhanced AI-Driven Intent Recognition Process

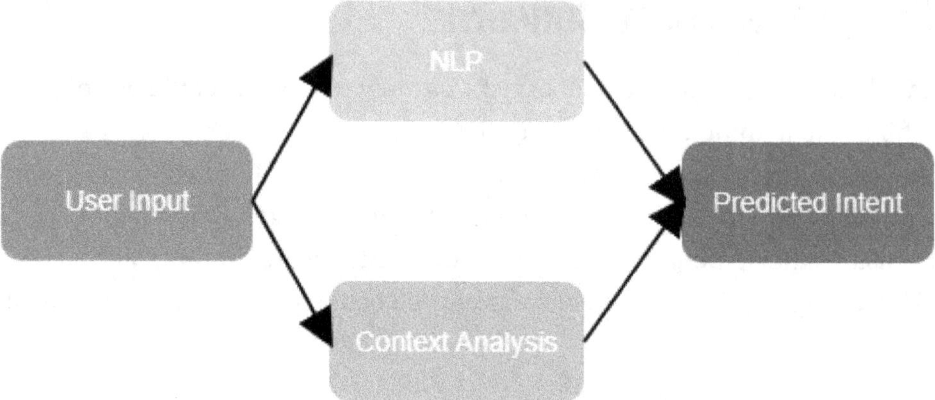

Figure 2-2. *Enhanced AI-Driven Intent Recognition Process*

Benefits of AI-Driven Intent Recognition

- **Enhanced Precision:** AI systems minimize errors in interpreting user intent, ensuring accurate execution. This precision reduces the risk of misconfigurations and enhances user satisfaction.

- **Proactive Adaptation:** By anticipating user needs, AI enables networks to adapt proactively, improving performance and user satisfaction. This anticipatory capability allows networks to remain agile and responsive.

- **Simplified Interaction:** Natural language interfaces make it easier for nontechnical users to interact with the network. This democratizes network management, allowing a broader range of users to define their networking needs.

Conclusion

Intent-based networking (IBN) marks a transformative shift in network management, driven by adaptive AI agents that interpret user intent, automate configurations, and ensure continuous alignment with desired outcomes. By leveraging AI capabilities like natural language processing and predictive analytics, IBN simplifies operations, enhances scalability, and proactively adapts to changing conditions.

Despite challenges such as intent ambiguity, legacy systems, and security concerns, advancements in AI and distributed architectures are enabling organizations to overcome these barriers. Adaptive AI agents are not only redefining how networks are managed but also bridging the gap between technical complexity and user needs.

As networks grow more dynamic, the role of AI in IBN will expand further, offering intelligent, scalable, and user-focused solutions. The next chapter will explore real-world implementations of AI agents, showcasing their practical applications and benefits across industries.

CHAPTER 3

Federated Learning for Collaborative Network Optimization

In today's interconnected world, where data is generated at an unprecedented scale across distributed systems, there is a growing need for efficient, collaborative, and privacy-preserving machine learning techniques. Federated learning (FL) has emerged as a groundbreaking paradigm, enabling multiple devices or systems to collaboratively train machine learning models without transferring sensitive raw data to a central location. This decentralized approach not only ensures privacy but also reduces the computational and communication overhead associated with traditional centralized learning models.

> *Federated learning is revolutionizing how networks collaborate and optimize themselves while maintaining data privacy.*

Federated learning is particularly well-suited for network optimization, where data is inherently distributed across endpoints such as routers, edge devices, and mobile nodes. In this chapter, we will explore the fundamental concepts of federated learning, delve into its applications in network optimization, examine privacy considerations, and analyze real-world case studies that demonstrate its transformative potential.

Fundamentals of Federated Learning

Federated learning represents a shift from traditional machine learning paradigms, focusing on decentralization and collaboration. It allows distributed nodes to train a shared model collaboratively while keeping their data localized. This approach is especially relevant in scenarios where data privacy, communication efficiency, and scalability are critical.

© Het Mehta 2026
H. Mehta, *AI Agents for Secure and Software-Defined Networking,*
https://doi.org/10.1007/979-8-8688-2358-9_3

Key Principles of Federated Learning

Federated learning operates on several foundational principles that distinguish it from traditional machine learning:

- **Decentralization:** Unlike centralized learning, which requires aggregating all data in one location, federated learning distributes the training process across multiple nodes. The model learns from diverse data sources without compromising their privacy.

- **Privacy Preservation:** Raw data never leaves the local devices. Instead, only model updates—such as gradients or parameters— are shared with a central server, significantly reducing the risk of exposing sensitive information.

- **Scalability:** FL is designed to scale across millions of devices, making it ideal for large-scale applications such as mobile networks, IoT systems, and cloud-edge ecosystems.

- **Collaborative Intelligence:** By leveraging diverse datasets from distributed nodes, federated learning creates robust and generalized models that outperform those trained on isolated datasets.

Federated Learning Workflow

The workflow of federated learning involves a series of iterative steps that enable nodes to collaboratively train a global model. These steps include

- **Model Initialization:** A global machine learning model is initialized by the central server and distributed to all participating nodes. This model serves as the starting point for training.

- **Local Training:** Each node trains the model on its local dataset. During this phase, the model learns patterns and features specific to the node's data.

- **Model Updates:** After local training, each node generates updates (e.g., gradients or parameters) that represent the knowledge gained during training.

- **Aggregation:** The central server collects updates from all participating nodes and aggregates them to create an updated global model. Techniques such as Federated Averaging (FedAvg) are commonly used for aggregation.

- **Model Distribution:** The updated global model is redistributed to the nodes for the next iteration of training.

- **Convergence:** The process continues iteratively until the global model achieves the desired level of accuracy and performance.

This iterative and collaborative process ensures that the global model benefits from diverse datasets while maintaining data privacy.

Federated Learning Workflow

This figure illustrates the workflow of federated learning, showing how a global model is distributed to multiple nodes, locally trained on their data, and then updated through aggregation at a central server. The iterative process enables collaborative learning without sharing raw data.

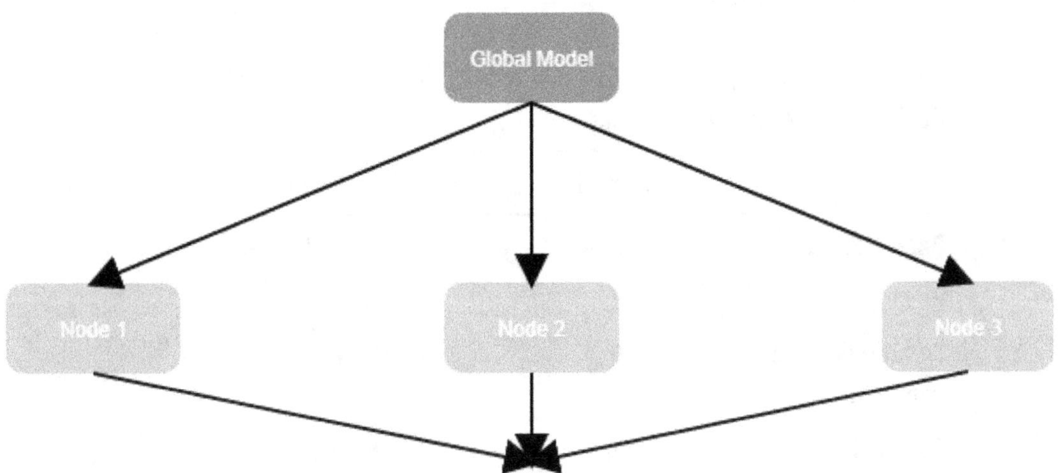

Figure 3-1. *Federated Learning Workflow*

Advantages of Federated Learning

Federated learning offers several advantages over traditional machine learning approaches:

- **Data Privacy:** By keeping data localized, FL minimizes the risk of data breaches and ensures compliance with privacy regulations such as GDPR.

- **Reduced Communication Overhead:** Since only model updates are transmitted, FL significantly reduces the bandwidth required for data transfer.

- **Enhanced Model Generalization:** By leveraging diverse datasets from multiple nodes, FL creates models that are robust and adaptable to various scenarios.

- **Scalability:** FL can scale to millions of devices, making it suitable for large-scale applications in healthcare, finance, and telecommunications.

Table 3-1. *Comparing Key Features of Centralized and Federated Learning Models*

Feature	Centralized Learning	Federated Learning
Data Storage	Centralized	Decentralized
Privacy	Low	High
Communication Overhead	High	Low
Scalability	Limited	High
Model Generalization	Moderate	High

Applications in Network Optimization

Federated learning has transformative applications in network optimization, enabling collaborative intelligence across distributed systems. Networks are inherently decentralized, with data generated at multiple endpoints such as routers, switches, and edge devices. FL leverages this distributed nature to optimize network performance, security, and resource allocation.

Traffic Prediction and Load Balancing

One of the most critical challenges in network optimization is predicting traffic patterns and balancing loads across the network. Federated learning enables nodes to collaboratively train models that forecast traffic demand in real time. By analyzing data from multiple endpoints, FL provides insights into peak usage times, congestion points, and traffic flow patterns.

For example, in a mobile network, FL can predict user demand in different geographical regions, enabling dynamic allocation of bandwidth and computing resources. This proactive approach reduces latency, prevents congestion, and ensures a seamless user experience.

Anomaly Detection

Anomaly detection is essential for maintaining the security and reliability of networks. Federated learning enhances anomaly detection by enabling nodes to collaboratively train models that identify irregularities in network behavior. This approach is particularly effective for detecting distributed anomalies, such as coordinated cyberattacks or unusual traffic patterns.

For instance, in a cloud network, FL can detect anomalies such as sudden spikes in data usage, unauthorized access attempts, or unusual data transfer rates. By leveraging data from multiple nodes, FL creates a holistic view of network behavior, improving the accuracy of anomaly detection.

Quality of Service (QoS) Optimization

Quality of Service (QoS) metrics, such as latency, bandwidth, and packet loss, are critical for ensuring optimal network performance. Federated learning enables the optimization of QoS parameters by analyzing data from multiple endpoints. This ensures that the network meets the performance requirements of various applications and users.

For example, in a video streaming network, FL can optimize QoS parameters to ensure smooth playback, minimal buffering, and high-quality resolution. By analyzing data from user devices, FL adapts the network to meet changing demands in real time.

Privacy Considerations in Federated Learning

While federated learning is inherently designed to preserve privacy, it is not immune to privacy risks. Adversaries can exploit shared model updates to infer sensitive information about local datasets. To address these risks, several privacy-enhancing techniques have been developed.

Differential Privacy

Differential privacy is a mathematical framework that ensures the privacy of individual data points by adding noise to the shared model updates. This noise prevents adversaries from inferring sensitive information while preserving the overall utility of the model.

For example, in a healthcare network, differential privacy can be used to protect patient data during federated learning. By adding noise to model updates, FL ensures that sensitive information, such as medical diagnoses, cannot be inferred by the central server.

Secure Aggregation

Secure aggregation techniques ensure that the central server can only access aggregated model updates, not individual contributions. This prevents the server from inferring sensitive information about individual nodes.

For instance, in a financial network, secure aggregation can protect transaction data during federated learning. By encrypting model updates, FL ensures that sensitive financial information remains confidential.

Federated Adversarial Training

Federated adversarial training enhances the robustness of the shared model against adversarial attacks. By incorporating adversarial examples during training, the model becomes more resilient to malicious nodes.

For example, in an IoT network, federated adversarial training can protect the shared model from compromised devices that attempt to inject malicious updates.

Techniques in Federated Learning

This diagram demonstrates the various privacy-preserving techniques used in federated learning, including differential privacy (noise addition), secure aggregation (encryption), and federated adversarial training (adversarial examples). These techniques enhance the security and robustness of the federated learning process.

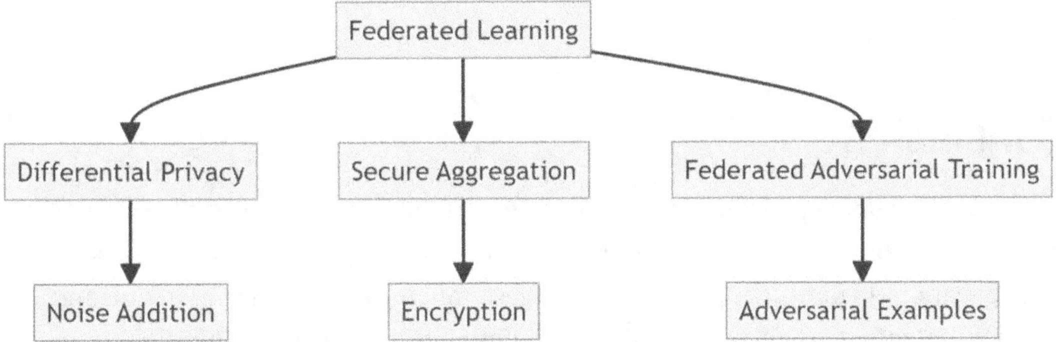

Figure 3-2. *Illustrating Various Privacy Techniques Used in Federated Learning*

Case Studies in Federated Learning

Federated learning is already transforming industries. The following case studies highlight its practical applications, starting with Google's innovative use in mobile devices.

Google's Federated Learning for Mobile Devices

Google has successfully implemented federated learning in its Gboard application, enabling personalized language models to be trained on user devices. This approach preserves user privacy while improving the accuracy of predictive text suggestions. By leveraging data from millions of devices, FL creates a robust language model that adapts to diverse user preferences.

Federated Learning in IoT Networks

In IoT networks, federated learning has been used to optimize device communication and energy consumption. For example, FL can train models to predict device failures, enabling proactive maintenance and reducing downtime. This approach improves the reliability and efficiency of IoT networks.

Federated Learning for Smart Cities

Smart cities generate vast amounts of data from sensors, cameras, and other devices. Federated learning enables these devices to collaboratively train models for applications such as traffic management, energy optimization, and public safety. By preserving data privacy, FL ensures that sensitive information, such as surveillance footage, remains secure.

Conclusion

Federated learning represents a paradigm shift in collaborative machine learning, enabling distributed systems to optimize themselves while preserving data privacy. Its applications in network optimization are vast, ranging from traffic prediction and anomaly detection to QoS optimization and smart city management. By addressing privacy challenges and leveraging the power of collaboration, federated learning paves the way for the next generation of intelligent networks.

As federated learning continues to evolve, its potential to transform industries such as telecommunications, healthcare, and finance will only grow. By embracing this innovative paradigm, organizations can unlock new levels of efficiency, security, and performance in their networks.

CHAPTER 4

AI Agents for Cross-Layer Network Orchestration

Modern networks are complex, multi-layered systems that require efficient coordination across various layers to ensure optimal performance. Traditionally, network orchestration relied on static policies and manual configurations, which often failed to meet the dynamic demands of today's applications. Enter AI agents—autonomous, intelligent systems capable of analyzing, learning, and optimizing network behavior across multiple layers.

> *AI agents are redefining how we approach network orchestration by enabling intelligent, cross-layer collaboration to optimize performance, scalability, and user experience.*

Cross-layer network orchestration refers to the coordinated management of different network layers, such as the physical, data link, network, and application layers, to achieve end-to-end optimization. By integrating AI agents into this process, networks can dynamically adapt to changing conditions, predict potential bottlenecks, and deliver seamless user experiences. This chapter explores the architecture of cross-layer orchestration, the role of AI techniques, performance metrics for evaluation, and the challenges associated with implementing AI-driven orchestration.

Cross-Layer Architecture Overview

Cross-layer architecture is a design paradigm that breaks the traditional siloed approach to network management. Instead of managing each layer independently, cross-layer architecture enables layers to share information and collaborate for holistic optimization. This section provides an overview of the architecture, its evolution, and key components.

27

© Het Mehta 2026
H. Mehta, *AI Agents for Secure and Software-Defined Networking*,
https://doi.org/10.1007/979-8-8688-2358-9_4

Traditional Layered Model vs. Cross-Layer Design

The traditional OSI (Open Systems Interconnection) model organizes networks into independent layers, each responsible for specific functions. While this modularity simplifies design and troubleshooting, it limits the ability to optimize performance across layers. For example, the network layer may not be aware of application-specific requirements, leading to suboptimal routing decisions.

In contrast, cross-layer design allows layers to exchange information and make collaborative decisions. For instance, the application layer can communicate its Quality of Service (QoS) requirements to the network layer, enabling intelligent resource allocation. Similarly, the physical layer can provide real-time feedback on signal strength and interference to the upper layers, ensuring efficient data transmission.

Example:

Consider a video streaming application where the application layer requires low latency and high bandwidth for smooth playback. In a traditional layered model, the network layer might not prioritize this traffic, leading to buffering and degraded quality. However, in a cross-layer design, the application layer can directly inform the network layer of its requirements, ensuring optimized routing and resource allocation.

Key Components of Cross-Layer Architecture

A cross-layer architecture typically consists of the following components:

- **Information Exchange Mechanisms:** Protocols or APIs that enable layers to share data such as bandwidth requirements, delay statistics, and energy consumption. These mechanisms form the backbone of cross-layer communication.

- **Orchestration Engine:** A centralized or distributed system that processes information from multiple layers and makes optimization decisions. The orchestration engine is often powered by AI algorithms that analyze data and predict network behavior.

- **Feedback Loops:** Mechanisms for monitoring and adjusting network behavior in real time based on performance metrics. Feedback loops ensure that the network can respond dynamically to changing conditions, such as traffic surges or hardware failures.

Cross-Layer Architecture

This figure provides a detailed representation of the cross-layer architecture, showcasing how different network layers—application, transport, network, and physical—interact and exchange information to achieve holistic optimization. Unlike the traditional siloed approach, this architecture enables seamless communication between layers, allowing for dynamic adjustments based on real-time network conditions. For example, the application layer can share QoS requirements with the network layer, while the physical layer can provide feedback on signal strength and interference, all working together to ensure efficient resource management and enhanced network performance.

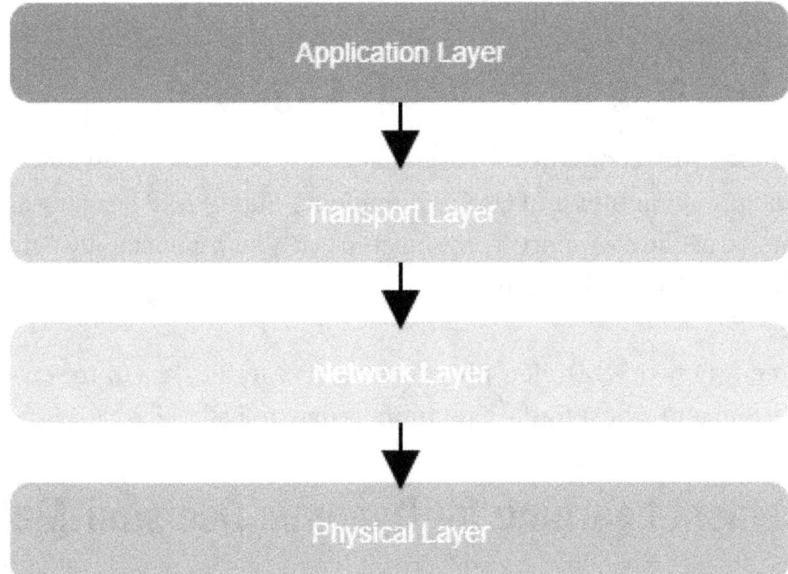

Figure 4-1. *Cross-Layer Architecture*

Benefits of Cross-Layer Design

- **Improved Performance:** By sharing information across layers, networks can optimize resource allocation, reduce latency, and enhance throughput.

- **Enhanced Flexibility:** Cross-layer design adapts to changing network conditions, ensuring consistent performance even under varying loads.

- **Better User Experience:** Applications benefit from tailored network configurations, resulting in smoother performance and higher user satisfaction.

AI Techniques for Orchestration

AI agents are at the heart of cross-layer network orchestration. By leveraging techniques such as machine learning, reinforcement learning, and deep learning, AI agents can analyze complex network data, predict future states, and make intelligent decisions.

Machine Learning for Pattern Recognition

Machine learning (ML) techniques are used to identify patterns in network data, such as traffic flow, congestion points, and QoS requirements. Supervised learning algorithms, such as decision trees and support vector machines (SVMs), can classify network states and recommend optimal configurations.

Example:

In a smart city network, ML algorithms can analyze traffic data from sensors to predict congestion and adjust traffic light timings dynamically.

Reinforcement Learning for Dynamic Decision-Making

Reinforcement learning (RL) is particularly well-suited for dynamic network environments. RL agents learn by interacting with the network and receiving feedback in the form of rewards or penalties. For example, an RL agent can optimize routing by learning which paths minimize latency and maximize throughput.

Real-World Application

In 5G networks, RL agents are used to manage spectrum allocation, ensuring that resources are distributed efficiently among users and applications.

Deep Learning for Complex Data

Deep learning models, such as convolutional neural networks (CNNs) and recurrent neural networks (RNNs), excel at processing complex, high-dimensional data. In cross-layer orchestration, deep learning can be used to predict QoS metrics, detect anomalies, and forecast network traffic.

Example:

Deep learning models can analyze video streaming data to predict bandwidth requirements and adjust network configurations in real time.

***Table 4-1.** Comparison of Various AI Techniques and Their Applications in Cross-Layer Orchestration*

AI Technique	Application in Orchestration	Strengths	Limitations
Machine Learning	Traffic classification, anomaly detection	Fast pattern recognition	Requires labeled data
Reinforcement Learning	Routing, resource allocation	Adapts to dynamic environments	High computational cost
Deep Learning	QoS prediction, fault detection	Handles complex data structures	Requires large datasets

Hybrid AI Approaches

Combining multiple AI techniques can enhance the effectiveness of cross-layer orchestration. For example, machine learning can be used for initial pattern recognition, while reinforcement learning fine-tunes decisions based on real-time feedback.

Performance Metrics and Evaluation

Evaluating the effectiveness of AI-driven cross-layer orchestration requires robust performance metrics. These metrics ensure that the orchestration meets the desired objectives, such as low latency, high throughput, and energy efficiency.

Key Performance Metrics

- **Latency:** The time taken for data to travel from source to destination. AI agents aim to minimize latency by optimizing routing and resource allocation.

- **Throughput:** The amount of data transmitted successfully over a network. High throughput indicates efficient resource utilization.

- **Packet Loss:** The percentage of data packets lost during transmission. AI agents reduce packet loss by predicting and mitigating network congestion.

- **Energy Efficiency:** The amount of energy consumed per unit of data transmission. AI agents optimize energy usage by dynamically adjusting power levels and routing paths.

Evaluation Framework

An effective evaluation framework for cross-layer orchestration includes

- **Simulation Environments:** Tools such as NS3 or OMNeT++ for testing AI algorithms in virtual network scenarios

- **Real-World Deployments:** Field trials to validate the performance of AI agents in operational networks

- **Benchmarking:** Comparing AI-driven orchestration with traditional approaches using standardized metrics

Challenges in Cross-Layer Orchestration

Despite its potential, AI-driven cross-layer orchestration faces several challenges.

Scalability

As networks grow in size and complexity, scaling AI agents to manage millions of devices becomes a significant challenge. Distributed AI frameworks, such as federated learning, can address this issue by enabling collaborative learning across devices.

Data Privacy

Sharing data across layers raises privacy concerns, especially in sensitive applications such as healthcare and finance. Techniques such as differential privacy and secure aggregation can mitigate these risks.

Real-Time Processing

Real-time orchestration requires AI agents to process large volumes of data with minimal latency. Edge computing can help by offloading computation to edge devices, reducing the burden on centralized servers.

Conclusion

AI agents for cross-layer network orchestration are revolutionizing how networks are managed, enabling intelligent, collaborative decision-making that adapts to dynamic conditions and optimizes performance. As networks grow increasingly complex with the rise of 5G, IoT, and edge computing, AI-driven orchestration provides the scalability and efficiency needed to handle these demands. By analyzing vast amounts of data in real time, AI agents improve resource utilization, reduce latency, and enhance energy efficiency, contributing to greener and more sustainable networks.

However, challenges such as scalability, privacy, and real-time processing must be addressed for widespread adoption. Techniques like federated learning and edge computing will play a critical role in overcoming these hurdles, enabling networks to respond faster and more securely. The future of AI-driven orchestration holds exciting possibilities, including self-optimizing networks, self-healing systems, and architectures that adapt seamlessly to new demands.

In summary, AI agents are not just a technological advancement but a necessity for the future of networking. By enabling smarter, faster, and more adaptable networks, they ensure that the growing complexity of modern systems can be managed effectively, paving the way for a new era of intelligent network management.

CHAPTER 5

Digital Twins for AI-Driven Network Simulations

The last decade has witnessed an unprecedented surge in the complexity, scale, and dynamism of digital networks. As organizations embrace cloud computing, 5G, IoT, and artificial intelligence, their networks have evolved into intricate ecosystems with thousands of interconnected devices and services. Managing, optimizing, and securing these networks using traditional approaches has become increasingly challenging. The stakes are high: downtime or security breaches can lead to significant financial losses, reputational damage, and even threats to public safety.

Enter the digital twin—a breakthrough concept that is transforming how networks are designed, managed, and evolved. A digital twin is more than just a simulation; it is a living, breathing digital replica of a physical network. By continuously synchronizing with real-world data, a digital twin provides a powerful platform for real-time monitoring, predictive analytics, scenario testing, and automated optimization.

This chapter explores the transformative role of digital twins in AI-driven network simulations. We will delve into the fundamental concepts, examine the technical architecture, and present real-world use cases. We will also discuss future trends, emerging challenges, and the many benefits digital twins bring to network management. Throughout the chapter, diagrams, tables, and practical examples will help illuminate this innovative field.

Concept of Digital Twins in Networking

To fully appreciate the impact of digital twins on networking, it is important to start with a clear understanding of what they are and the foundational principles that guide their design and use. The next section provides a detailed definition and explores the core concepts behind digital twins in the context of modern networks.

© Het Mehta 2026
H. Mehta, *AI Agents for Secure and Software-Defined Networking,*
https://doi.org/10.1007/979-8-8688-2358-9_5

Definition and Core Principles

A digital twin is a real-time, virtual representation of a physical system—such as a network—that mirrors its state, behavior, and performance. Unlike traditional network models, which are often static and quickly outdated, a digital twin is dynamic and interactive. It ingests live data from the actual network, continuously updating itself to reflect the current reality. This connection allows for a closed feedback loop: insights and optimizations discovered in the digital realm can be applied directly to the physical network, and vice versa.

Key principles that distinguish digital twins in networking include

- **Synchronization:** Digital twins are continuously updated with real-world data, ensuring their accuracy and relevance.

- **Interactivity:** Users can interact with the digital twin to test scenarios, make changes, and observe outcomes without risk to the live network.

- **Lifecycle Support:** Digital twins support the entire lifecycle of network assets, from design and deployment to operation and decommissioning.

- **Predictive Power:** By leveraging AI and analytics, digital twins can forecast network behavior, detect anomalies, and recommend proactive interventions.

Transitioning from static models to dynamic digital twins marks a significant leap in how networks are understood and managed.

Key Elements of a Digital Twin Network

For a digital twin to deliver on its promise, several essential components must work in harmony:

- **Data Collection:** Sensors, logs, and telemetry systems gather real-time information about network devices, traffic flows, configurations, and environmental conditions.

- **Mapping and Modeling:** The collected data is used to construct a comprehensive virtual map of the network, including topology, device characteristics, and logical relationships.

- **Simulation Engine:** Sophisticated algorithms and AI models simulate network behavior under various conditions, enabling scenario testing and optimization.

- **Visualization and Analytics:** Dashboards and analytics tools provide intuitive visualizations of network status, performance metrics, and predictive insights.

- **Control Interfaces:** Secure APIs and automation tools allow changes and optimizations identified in the digital twin to be safely implemented in the real network.

These elements create a robust feedback system, where the digital twin not only mirrors the physical network but also actively enhances its operation.

Architectural Overview

A well-architected digital twin platform is built in modular layers, each with a distinct role. The following diagram illustrates a typical architecture:

Enhanced Digital Twin Network Architecture

The "Digital Twin Network Architecture" diagram presents a layered system where physical network components like routers and IoT devices feed data through sensors and APIs into an AI-powered modeling engine. This engine creates digital replicas and optimizes the network. A simulation environment allows scenario testing, while an analytics dashboard provides monitoring and insights. Finally, a feedback loop with control and automation ensures continuous validation and adjustment, enabling smart and adaptive network management.

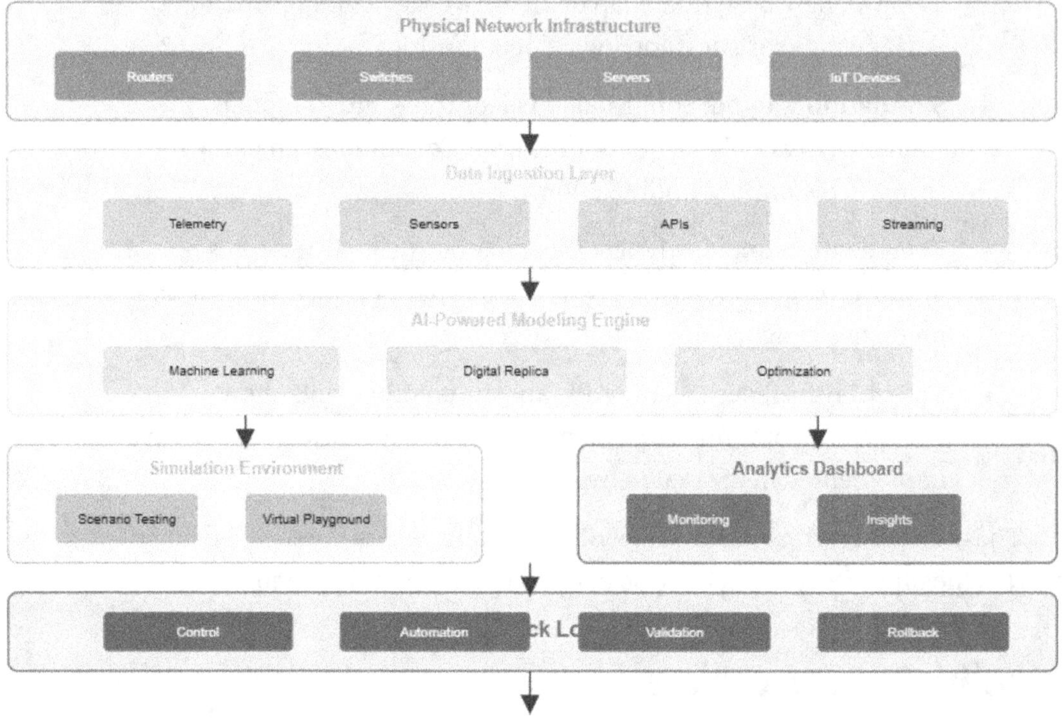

Figure 5-1. *Enhanced Architecture of a Digital Twin Network*

Evolution and Industry Adoption

Digital twins first emerged in manufacturing and aerospace, where they were used to monitor and optimize physical assets. In the networking domain, digital twins have evolved from simple emulation tools to sophisticated, AI-powered platforms. Today, they are being adopted by telecom operators, cloud providers, and large enterprises to address the growing need for proactive, data-driven management of complex network environments.

Simulation Scenarios and Use Cases

Overview

Digital twins unlock a wide variety of simulation scenarios, each designed to address specific challenges in network management. Their ability to mirror real-world conditions, test hypotheses, and predict outcomes makes them invaluable for planning, optimization, and risk mitigation.

Network Optimization and Capacity Planning

One of the primary uses of digital twins is to optimize network performance and plan for future growth. By simulating different traffic patterns, routing protocols, and resource allocation strategies, operators can identify bottlenecks and inefficiencies before they impact users.

Example: A global telecom operator plans to roll out a new 5G service in a metropolitan area. Using a digital twin, the operator simulates user demand, signal propagation, and interference across various antenna placements. The simulation reveals potential coverage gaps and congestion points, enabling the operator to adjust the network design before any physical deployment. This proactive approach saves time, reduces costs, and ensures a better user experience from day one. The AI-Powered Modeling Engine can also employ reinforcement learning algorithms to explore optimal antenna configurations and deep learning models to predict signal propagation patterns, enabling this predictive planning capability.

Cybersecurity Testing and Threat Modeling

Security is paramount in modern networks. Digital twins provide a safe, isolated environment for simulating cyberattacks and testing defense mechanisms. Operators can model DDoS attacks, malware outbreaks, and insider threats, observing how the network responds and identifying weaknesses.

Example: A financial institution creates a digital twin of its core network to simulate a ransomware attack. The simulation reveals that several critical servers lack proper segmentation and backup procedures. As a result, the organization strengthens its defenses and updates its incident response plan, reducing the risk of real-world breaches.

Predictive Maintenance and Fault Detection

Digital twins excel at predictive maintenance by continuously monitoring network health and analyzing trends. AI models within the twin can forecast failures based on historical and live data, enabling proactive interventions.

Example: A cloud service provider uses a digital twin to monitor thousands of servers across multiple data centers. The twin detects subtle signs of impending disk failure—such as increased error rates and latency—allowing the provider to replace hardware before any data loss occurs. This approach minimizes downtime and enhances service reliability.

Training and Education

Digital twins offer realistic, hands-on training environments for network engineers and security professionals. Trainees can practice troubleshooting, experiment with new configurations, and simulate attacks without risking production systems.

Example: A university integrates a digital twin into its cybersecurity curriculum. Students use the twin to simulate network attacks, apply defensive measures, and analyze outcomes. This immersive experience accelerates learning and prepares graduates for real-world challenges.

Disaster Recovery and Resilience Testing

Ensuring network resilience is critical for business continuity. Digital twins enable operators to simulate disasters—such as power outages, hardware failures, or cyber incidents—and test recovery procedures in a controlled environment.

Example: A multinational corporation uses a digital twin to simulate a complete data center outage. The simulation tests failover mechanisms, backup systems, and restoration processes, ensuring that the organization can recover quickly and maintain operations in the face of real disasters.

Smart City and Infrastructure Applications

Digital twins are increasingly being used in smart city initiatives to simulate and optimize urban infrastructure. By integrating data from transportation, utilities, and public services, digital twins help planners design more efficient, resilient, and sustainable cities.

Example: A city government employs a digital twin to analyze traffic flow, energy consumption, and environmental impact. The simulation identifies opportunities to reduce congestion, optimize public transport routes, and lower energy usage, contributing to a higher quality of life for residents.

Future Directions in Digital Twin Technology

As digital twin technology continues to mature, its potential applications and capabilities are rapidly expanding. The future promises even greater innovation as digital twins intersect with other cutting-edge developments in the technology landscape. The following sections explore how these advancements will shape the next generation of digital twin solutions, beginning with their integration alongside emerging technologies.

Integration with Emerging Technologies

The future of digital twins is closely tied to the evolution of other cutting-edge technologies:

- **IoT and Edge Computing:** As networks become more distributed, digital twins will leverage IoT sensors and edge computing to collect granular, real-time data and process it locally. This reduces latency, enhances scalability, and enables faster decision-making.

- **5G and Beyond:** Digital twins will be indispensable for the deployment and optimization of next-generation wireless networks. They will allow operators to simulate complex radio environments, optimize antenna configurations, and predict network performance under diverse conditions.

Advanced AI and Machine Learning

AI will play an increasingly central role in digital twins, enabling

- **Deep Learning:** Enhanced pattern recognition and anomaly detection for more accurate predictions and faster troubleshooting

- **Reinforcement Learning:** Autonomous optimization of network configurations, adapting to changing conditions in real time

- **Natural Language Processing:** Voice-activated management and automated reporting, making digital twins more accessible and user-friendly

Interoperability and Standardization

As digital twins become more widespread, the need for industry-wide standards and interoperability frameworks will grow. These standards will facilitate seamless integration of digital twins across different platforms, vendors, and network domains, promoting innovation and reducing vendor lock-in.

Sustainability and Environmental Impact

Sustainability is becoming a key driver for digital twin adoption. By simulating energy consumption, emissions, and resource usage, digital twins enable operators to optimize for energy efficiency and reduce their environmental footprint. This is particularly important as organizations face increasing regulatory pressure to meet sustainability targets.

Example: A data center operator uses a digital twin to model power usage and cooling requirements. The simulation identifies opportunities to shift workloads, optimize cooling systems, and reduce overall energy consumption, supporting both cost savings and environmental goals.

Expansion into Smart Cities and Infrastructure

Digital twins are expanding into large-scale infrastructure and smart city applications. Urban planners use digital twins to simulate mobility patterns, traffic flows, and environmental changes, providing data-driven insights for better resource allocation and planning.

Example: A metropolitan region uses a digital twin to coordinate emergency response during a natural disaster. The twin integrates data from transportation, utilities, and public safety agencies, helping authorities allocate resources efficiently and communicate with residents in real time.

Benefits of Digital Twins for Network Management

Digital twins bring a host of transformative benefits to network management. The table below summarizes the key advantages.

Table 5-1. *Benefits of Digital Twins for Network Management*

Benefit	Description	Impact
Improved Decision-Making	Real-time insights and predictive analytics for proactive network decisions	Enhanced performance, reduced downtime, better resource utilization
Enhanced Security	Simulation of attack scenarios and vulnerability testing	Reduced risk of breaches, faster incident response
Cost Efficiency	Virtual testing reduces the need for physical trials and errors, ensuring faster deployment cycles	Significant savings in operational and capital expenditures
Increased Flexibility	Rapid scenario testing for quick adaptation to changing conditions	Greater agility in upgrades and configuration changes
Training and Skill Development	Safe, realistic environments for staff training	Improved workforce expertise, faster onboarding
Sustainability and Compliance	Monitoring and optimizing energy use for regulatory and environmental goals	Lower carbon footprint, regulatory compliance
Accelerated Innovation	Rapid prototyping and validation of new technologies and services	Faster time-to-market, competitive advantage
Business Continuity	Simulation of disaster recovery and resilience strategies	Higher availability, reduced impact of outages

Challenges and Best Practices

While digital twins offer remarkable advantages, their implementation is not without challenges. Organizations must consider the following:

- **Data Quality and Integration:** The accuracy of a digital twin depends on the quality and completeness of its data sources. Ensuring seamless integration and real-time updates is crucial.

- **Security and Privacy:** Digital twins must be protected from unauthorized access and cyber threats, as they can contain sensitive network information.

- **Scalability:** As networks grow, digital twins must scale efficiently to handle increasing data volumes and complexity.

- **Interoperability:** Avoiding vendor lock-in and ensuring compatibility with existing tools and platforms is essential for long-term success.

- **Change Management:** Introducing digital twins may require changes in organizational processes, roles, and skill sets. Effective training and change management are key.

Best Practices

1. Start with a pilot project focused on a specific network segment or use case.

2. Invest in robust data collection and integration infrastructure.

3. Prioritize security and privacy from the outset.

4. Foster cross-functional collaboration between IT, operations, and business units.

5. Continuously update and refine the digital twin as the network evolves.

Conclusion

Digital twins represent a paradigm shift in how networks are simulated, managed, and optimized. By creating dynamic, real-time virtual replicas of physical networks, digital twins empower organizations to anticipate problems, optimize resources, and innovate with confidence.

This chapter has explored the foundational concepts, technical architecture, and real-world applications of digital twins in networking. We have examined their role in network optimization, security, predictive maintenance, training, disaster recovery, and smart city development. We have also looked ahead to future trends, including AI integration, sustainability, and the expansion of digital twins into new domains.

The benefits of digital twins—ranging from improved decision-making and cost efficiency to enhanced security and sustainability—underscore their critical role in the future of network management. As networks continue to grow in complexity and importance, digital twins will become indispensable tools for building resilient, efficient, and secure digital infrastructures.

Organizations that embrace digital twins today will be well positioned to thrive in the connected world of tomorrow.

CHAPTER 6

Cognitive AI Agents for Proactive Network Healing

Modern digital networks are the lifeblood of today's interconnected society, powering industries as diverse as banking, healthcare, manufacturing, logistics, and entertainment. As organizations scale their digital operations, networks grow in size and complexity, often spanning multiple cloud providers, data centers, and edge environments. This growth brings significant challenges: performance bottlenecks, unpredictable failures, cyber threats, and the constant need for rapid adaptation to shifting business requirements.

Traditional network management, with its reliance on manual configuration, static rules, and reactive troubleshooting, is increasingly unable to keep pace. Outages, slowdowns, and security breaches can have devastating consequences, including lost revenue, reputational harm, and even threats to public safety. To meet these challenges, organizations are turning to a new generation of intelligent, self-managing networks driven by cognitive AI agents.

Cognitive AI agents are inspired by human cognition—they perceive, learn, reason, and act autonomously. These agents can proactively identify and resolve network problems, often before users notice any impact. By leveraging advanced artificial intelligence, machine learning, and real-time analytics, cognitive agents enable networks to self-heal, optimize their own operations, and deliver continuous, high-quality service with minimal human intervention.

This chapter explores the core principles of cognitive networking, the AI techniques that underpin proactive healing, and real-world applications that demonstrate measurable value. We also examine the challenges and best practices for deploying cognitive healing solutions, equipping readers to harness the full potential of this transformative technology.

© Het Mehta 2026
H. Mehta, *AI Agents for Secure and Software-Defined Networking,*
https://doi.org/10.1007/979-8-8688-2358-9_6

Principles of Cognitive Networking

Cognitive networking represents a paradigm shift in how networks are designed, managed, and evolved. Rather than relying solely on static policies or manual oversight, cognitive networks are capable of understanding their environment, learning from experience, and adapting their behavior in real time.

What Makes a Network "Cognitive"?

A cognitive network is distinguished by several fundamental attributes:

- **Awareness:** The network senses a wide range of internal and external factors, such as device health, traffic patterns, user behavior, and environmental conditions.

- **Learning:** It continuously analyzes historical and real-time data to recognize patterns and improve its responses over time.

- **Reasoning:** The network uses logic and inference to diagnose problems, predict outcomes, and select optimal actions.

- **Adaptation:** It dynamically adjusts its configuration and policies in response to changing conditions or business goals.

- **Autonomy:** The network acts independently, minimizing the need for human intervention and reducing operational overhead.

These attributes enable cognitive networks to move beyond traditional, reactive management and toward proactive, preventative operations.

The Cognitive Networking Loop

At the heart of cognitive networking is a continuous feedback process, often modeled as the **Observe–Orient–Decide–Act (OODA) loop**. This loop empowers the network to

1. **Observe:** Gather data from devices, applications, users, and the environment.

2. **Orient:** Analyze and contextualize the data to understand the current state and trends.

3. **Decide:** Use AI-driven reasoning to determine the best course of action.

4. **Act:** Implement changes, such as rerouting traffic, updating configurations, or applying patches.

5. **Learn:** Evaluate the outcomes of actions and refine future responses accordingly.

This loop is perpetual, allowing the network to evolve and improve its healing and optimization capabilities with each iteration.

Cognitive Network Healing Loop Diagram

The Cognitive Network Healing Loop diagram illustrates a continuous process where the network detects issues, analyzes root causes using cognitive techniques, and applies corrective actions automatically. This loop includes monitoring, diagnosis, decision-making, and healing phases, enabling the network to self-repair and maintain optimal performance with minimal human intervention.

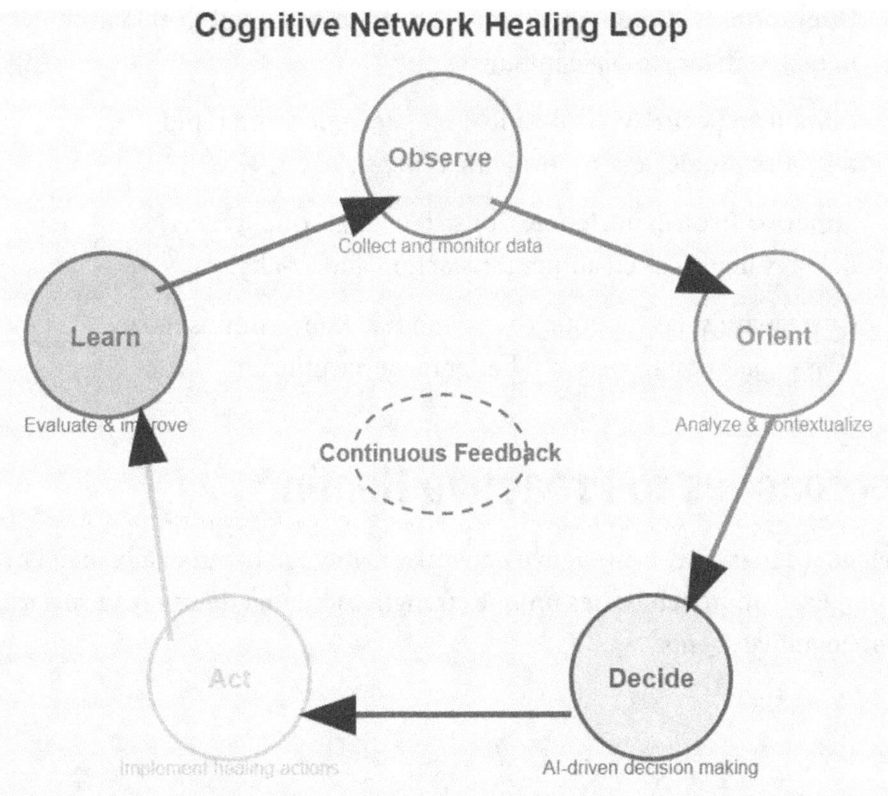

Figure 6-1. *The Cognitive Network Healing Loop*

Cognitive Networking in Practice

Cognitive networking is not just theoretical—it is increasingly being adopted in enterprise, telecom, and cloud environments. For example, major cloud providers use cognitive principles to automate traffic management and incident response, while telecom operators apply them to optimize service quality and minimize downtime.

Benefits of Cognitive Networking

By embedding cognitive intelligence into network management, organizations can realize

1. **Faster Incident Resolution:** Automated detection and remediation of faults reduce mean time to repair (MTTR).

2. **Improved Reliability:** Proactive healing prevents minor issues from growing into major outages.

3. **Operational Efficiency:** Automation reduces manual workload, freeing staff for strategic initiatives.

4. **Enhanced Security:** Real-time threat detection and rapid response protect against evolving cyber risks.

5. **Superior User Experience:** Consistent, high-quality service delivery increases customer satisfaction and loyalty.

6. **Cost Savings:** Fewer outages and manual interventions mean lower operational costs and better resource utilization.

AI Approaches to Proactive Healing

The intelligence that powers proactive network healing is a blend of advanced AI techniques. Each approach brings unique strengths, and together they form a robust toolkit for cognitive agents.

Machine Learning for Anomaly Detection

Machine learning (ML) is the foundation for recognizing abnormal patterns in network data. By training on vast datasets of logs, metrics, and historical incidents, ML models can

- Identify subtle deviations from normal behavior (e.g., unusual latency or packet loss)

- Detect early signs of hardware degradation, misconfigurations, or security threats

- Continuously adapt to evolving network conditions

Types of ML Used

1. **Supervised Learning:** Uses labeled data (e.g., "failure" vs. "healthy") to detect known issues

2. **Unsupervised Learning:** Finds new or previously unknown anomalies without labels

3. **Semi-supervised Learning:** Combines both, maximizing accuracy where labeled data is limited

Deep Learning for Root Cause Analysis

Deep learning, with its multi-layered neural networks, excels at processing high-dimensional, complex data such as traffic flows, logs, and event sequences. It can

1. Map intricate relationships among diverse network signals

2. Pinpoint the root causes of cascading failures or performance issues

3. Identify correlations that humans or simpler algorithms might miss

Example: A deep neural network analyzes network logs and discovers that a recurring spike in latency correlates with periodic software updates on certain switches—enabling preemptive scheduling and resource allocation.

Reinforcement Learning for Automated Remediation

Reinforcement learning (RL) allows cognitive agents to learn optimal healing strategies through interaction and feedback. RL agents

1. Experiment with different actions (e.g., rerouting, restarting services)

2. Receive rewards or penalties based on the impact on network health

3. Develop policies that maximize long-term reliability and performance

Applications

- Dynamic traffic rerouting to avoid congestion or outages

- Automated configuration tuning for performance optimization

- Adaptive security policies responding to real-time threats

Knowledge-Based Systems and Reasoning

Knowledge-based systems encode expert rules, diagnostic workflows, and best practices. These systems

- Use logical reasoning to guide troubleshooting

- Supplement data-driven AI with human expertise

- Enable explainable, auditable decision-making

Hybrid Approach: Combining knowledge-based reasoning with ML and RL results in agents that are both adaptive and robust, leveraging the best of both worlds.

Natural Language Processing for Human–AI Collaboration

Natural language processing (NLP) bridges the gap between human operators and AI systems. NLP-powered interfaces can

- Interpret operator queries and commands

- Provide explanations, recommendations, and action plans in natural language

- Enable conversational troubleshooting and reporting

Example: An engineer asks, "Why is packet loss high in the Atlanta data center?" The agent analyzes telemetry, identifies a failing link, and responds with a detailed, actionable explanation.

Table 6-1. *AI Techniques in Proactive Network Healing*

AI Technique	Purpose	Example Application
Machine Learning	Anomaly detection, pattern recognition	Detecting traffic spikes
Deep Learning	Complex data analysis, root cause analysis	Identifying hardware failure precursors
Reinforcement Learning	Policy optimization, automated remediation	Dynamic rerouting during outages
Knowledge-Based Systems	Rule-based reasoning, best practices	Automated troubleshooting workflows
Natural Language Processing	Human–AI collaboration, explanations	Conversational troubleshooting assistants

Transitioning from theory to practice, let's examine how these techniques are deployed in real-world networks and the tangible benefits they deliver.

Real-World Applications and Outcomes

Cognitive AI agents are actively transforming network operations in a variety of industries. Their ability to proactively heal and optimize networks is delivering measurable improvements in reliability, security, and user experience.

Automated Fault Detection and Self-Healing

Large enterprises and service providers use cognitive agents to continuously monitor network health and automatically resolve issues. These agents

- Detect anomalies in real time

- Diagnose faults using AI-driven analysis

- Execute remediation steps such as restarting services, rerouting traffic, or updating configurations

Case Study: A global cloud provider deployed cognitive agents across its backbone. When a fiber cut disrupted a major link, the agent detected the anomaly, diagnosed the issue, and automatically rerouted traffic. End users experienced no noticeable downtime, and the incident was resolved before manual intervention was required.

Predictive Maintenance in Telecom Networks

Telecom operators face the challenge of maintaining thousands of hardware components. Cognitive agents use predictive analytics to

- Analyze historical failure data and sensor readings

- Forecast which components are likely to fail

- Schedule maintenance before failures occur, reducing unplanned outages

Example: A mobile operator implemented AI-driven predictive maintenance for its cell towers. The system analyzed equipment logs and weather data to predict failures, enabling preemptive repairs and reducing service interruptions by 30%.

Adaptive Security and Threat Mitigation

Cognitive agents play a vital role in network security by

- Detecting abnormal traffic patterns and user behaviors

- Identifying and mitigating cyber threats in real time

- Automatically updating security policies to counter new attack vectors

Case Study: A financial services firm faced repeated DDoS attacks. Cognitive agents detected the attacks, classified the threat vectors, and dynamically updated firewall rules to block malicious sources, preventing outages and safeguarding sensitive data.

Enhancing User Experience Through Proactive Optimization

User satisfaction depends on consistent network performance. Cognitive agents

- Monitor user engagement and network congestion

- Optimize resource allocation and load balancing

- Automatically adjust quality-of-service parameters to resolve issues before users are impacted

Example: A streaming media company used cognitive AI to monitor buffer rates and congestion. When increased buffering was detected in a region, the agent adjusted content delivery routes, improving video quality and reducing customer complaints.

Training and Augmenting Network Operations Teams

Cognitive agents also serve as intelligent assistants, helping human operators

- Analyze alerts and prioritize incidents

- Recommend and automate troubleshooting steps

- Reduce mean time to resolution (MTTR) and enable teams to focus on strategic work

Example: A managed service provider integrated a cognitive assistant in its operations center. The assistant analyzed alerts, suggested fixes, and automated common remediation, cutting MTTR by 40%.

Table 6-2. *Real-World Outcomes of Cognitive Healing*

Application Area	Achieved Outcome	Industry Example
Automated Fault Detection	Reduced downtime, faster recovery	Cloud infrastructure
Predictive Maintenance	Fewer outages, lower maintenance costs	Telecommunications
Adaptive Security	Improved threat response, minimized breaches	Financial services
User Experience Optimization	Higher customer satisfaction, reduced churn	Streaming media
Operations Team Augmentation	Increased efficiency, faster incident resolution	Managed IT services

Cognitive Healing in Multicloud and Hybrid Environments

As organizations move to multicloud and hybrid architectures, cognitive healing agents must operate across diverse platforms. This requires

1. Unified data collection and normalization across clouds

2. Policy orchestration that spans on-premises and cloud environments

3. AI models that adapt to heterogeneous infrastructure and workloads

Practical Example: A multinational enterprise uses cognitive agents to manage both its on-premises data centers and multiple public cloud environments. The agents coordinate failover, security, and optimization policies across all platforms, ensuring seamless service continuity and compliance.

Quantifying the Impact: Metrics and KPIs

Organizations measure the impact of cognitive healing using key performance indicators (KPIs) such as

- **Mean Time to Detect (MTTD):** Time taken to identify an issue

- **Mean Time to Repair (MTTR):** Time taken to resolve an issue

- **Service Availability:** Percentage of uptime achieved

- **User Satisfaction Scores:** Based on surveys and feedback

- **Incident Volume:** Number of outages or security breaches per period

Table 6-3. *Example KPIs Before and After Cognitive Healing Deployment*

KPI	Before Cognitive Healing	After Cognitive Healing	Improvement
MTTD (minutes)	25	5	80% faster
MTTR (minutes)	60	15	75% faster
Service Availability	99.5%	99.98%	+0.48% which is a 42 hr increment in the annual uptime
Incident Volume	12/month	3/month	75% reduction
User Satisfaction	7.2/10	9.1/10	+1.9 points

The successful deployment of cognitive healing agents demonstrates their value in diverse, real-world scenarios. However, achieving these benefits requires overcoming significant challenges.

Challenges in Cognitive Healing

While the promise of cognitive AI agents is immense, their implementation and operation are not without hurdles. Addressing these challenges is crucial for realizing the full potential of proactive network healing.

Data Quality and Integration

Cognitive agents depend on accurate, timely, and comprehensive data. Challenges include

- Integrating data from legacy systems, modern devices, and cloud platforms

- Dealing with incomplete, inconsistent, or noisy data

- Ensuring real-time data availability for timely decision-making

Best Practice: Invest in robust data pipelines, data normalization tools, and continuous data quality monitoring.

Model Interpretability and Trust

AI-driven decisions, especially from deep learning models, can appear as "black boxes." Operators may hesitate to trust automated actions without clear explanations.

Best Practice: Adopt explainable AI (XAI) techniques that provide transparent, auditable decision paths and clear justifications for actions taken.

Security and Privacy Concerns

Cognitive agents require access to sensitive network data, raising security and privacy issues:

- Protecting AI systems from tampering or unauthorized access

- Ensuring compliance with data protection regulations (e.g., GDPR, HIPAA)

- Implementing data anonymization and access controls

Best Practice: Apply strong security controls, regular audits, and privacy-by-design principles.

Scalability and Performance

As networks expand, cognitive healing solutions must

- Scale to handle increasing data volumes and device counts

- Maintain low-latency analysis and response

- Operate efficiently in distributed and cloud-native environments

Best Practice: Leverage distributed architectures, cloud-native deployment, and scalable AI frameworks.

Change Management and Human Factors

Transitioning to cognitive healing impacts people and processes:

- Operators need training to collaborate with AI systems.

- Clear escalation protocols must be established for cases requiring human intervention.

- Organizational culture may need to evolve to trust automation.

Best Practice: Invest in training, change management programs, and foster a culture of human–AI collaboration.

Table 6-4. *Key Challenges and Mitigation Strategies in Cognitive Healing*

Challenge	Description	Mitigation Strategy
Data Quality	Incomplete or inconsistent data impairs AI accuracy	Robust data collection/ integration
Model Interpretability	Opaque AI decisions hinder trust	Explainable AI, transparent reporting
Security and Privacy	Sensitive data access raises risks	Strong security, privacy-by-design
Scalability	Large networks strain resources	Distributed/cloud-native solutions
Change Management	Human factors affect adoption	Training, clear protocols, culture shift

Conclusion

Cognitive AI agents are at the forefront of a new era in network management—one defined by automation, intelligence, and resilience. By combining the power of awareness, learning, reasoning, and autonomy, cognitive networks can proactively detect, diagnose, and resolve issues, delivering unprecedented reliability and performance.

This chapter has explored the foundational principles of cognitive networking, the AI approaches that power proactive healing, and real-world deployments that showcase dramatic improvements in uptime, security, and user satisfaction. We have also examined the challenges and best practices for successful adoption.

As networks continue to grow in complexity and importance, cognitive AI agents will be indispensable for building self-healing, adaptive, and user-centric digital infrastructures. Organizations that embrace this technology will be prepared to deliver reliable services, respond to emerging threats, and thrive in the rapidly evolving digital landscape.

CHAPTER 7

AI-Powered Dynamic Policy Management in SDN

The digital transformation sweeping across industries has elevated the importance of agile, secure, and efficient networks. Software-Defined Networking (SDN) has emerged as a cornerstone of this evolution, enabling organizations to manage network resources with the flexibility and programmability that modern applications demand. At the heart of SDN is policy management—the set of rules and controls that determine how data flows, who can access resources, and how the network responds to threats or performance challenges.

Traditional policy management, however, is struggling to keep pace with the dynamic and complex nature of today's networks. Static rules and manual interventions are no longer sufficient to guarantee security, compliance, and optimal performance. The advent of artificial intelligence (AI) is now reshaping this landscape, transforming policy management from a rigid, reactive process into a dynamic, predictive, and autonomous capability.

This chapter explores the fundamentals of policy management in SDN, the AI techniques that enable real-time policy adaptation, and practical case studies that illustrate the transformative impact of AI-powered dynamic policy management. By understanding these concepts, network architects, engineers, and decision-makers can harness the full potential of SDN to support business agility, security, and innovation.

© Het Mehta 2026
H. Mehta, *AI Agents for Secure and Software-Defined Networking*,
https://doi.org/10.1007/979-8-8688-2358-9_7

Policy Management Fundamentals

Policy management is the foundation upon which modern networks are governed. In SDN, policies are high-level directives that control network behavior, ensuring the infrastructure aligns with business objectives, operational requirements, and regulatory constraints.

The Role of Policy in SDN

Policies in SDN can be broadly categorized into several types, each serving a distinct purpose:

- **Access Control Policies:** Define who or what can access specific network resources, often based on user identity, device type, or application.

- **Quality of Service (QoS) Policies:** Govern the allocation of bandwidth, prioritization of traffic, and management of latency for different applications or users.

- **Security Policies:** Specify how the network detects, prevents, and responds to threats, including firewall rules, segmentation, and encryption requirements.

- **Routing Policies:** Determine the paths that data packets take through the network, optimizing for performance, reliability, or cost.

These policies are typically expressed in high-level, intent-based language and are enforced by the SDN controller, which translates them into device-specific configurations.

The Policy Management Lifecycle

Effective policy management is not a static, one-time activity but a continuous lifecycle. This lifecycle ensures that the network remains responsive to evolving conditions, business priorities, and threat landscapes.

SDN Policy Management Lifecycle

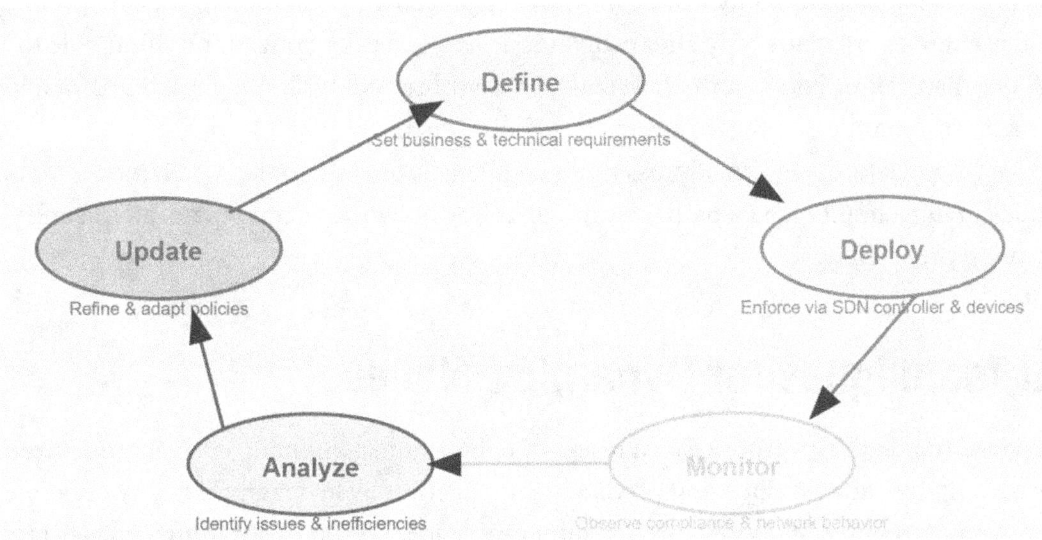

Figure 7-1. *SDN Policy Management Lifecycle*

The lifecycle begins with defining high-level policies aligned to organizational goals and technical constraints. Once defined, these policies are deployed across the network by the SDN controller, which translates them into actionable configurations. Continuous monitoring ensures that the network's behavior remains consistent with the policies. Analytics provide insights into deviations, inefficiencies, or emerging threats, prompting updates and refinements to policies as needed. This cycle repeats, creating a self-improving system that evolves with changing requirements.

Challenges of Traditional Policy Management

Despite the architectural advantages of SDN, traditional policy management methods are increasingly inadequate in the face of dynamic business needs and sophisticated cyber threats. Static rules are often too rigid to accommodate sudden shifts in network traffic, application demand, or security posture. Manual intervention is not only time-consuming but also prone to human error, especially as networks scale to thousands of devices and policies.

Complexity compounds the challenge, as administrators must juggle overlapping or conflicting policies across hybrid environments. Moreover, a reactive approach—where policy changes are made only after problems arise—leaves organizations vulnerable to service disruptions and security breaches. These shortcomings highlight the urgent need for a more dynamic, intelligent approach to policy management.

As organizations seek to address these challenges, the integration of AI into SDN policy management offers a path forward, enabling networks to anticipate, adapt, and respond proactively.

AI Techniques for Dynamic Policies

Artificial intelligence introduces a new era of policy management in SDN, characterized by automation, adaptability, and predictive intelligence. By leveraging AI, networks can sense environmental changes, predict outcomes, and autonomously adjust policies in real time.

Machine Learning for Policy Optimization

Machine learning (ML) plays a pivotal role in optimizing network policies. ML models are trained on vast datasets of network telemetry, historical policy outcomes, and user behavior. They can identify patterns—such as recurring congestion, anomalous traffic, or performance bottlenecks—and recommend or automatically implement policy changes to preempt issues.

For example, an ML-driven SDN controller might learn that specific times of day correspond with spikes in video traffic. In anticipation, it can adjust QoS policies to prioritize streaming applications, ensuring a seamless user experience without sacrificing bandwidth for critical business functions.

Reinforcement Learning for Adaptive Policy Control

Reinforcement learning (RL) extends the capabilities of policy management by enabling continuous adaptation. RL agents interact with the network environment, experimenting with different policy actions and receiving feedback in the form of rewards or penalties. Over time, the agent learns the optimal set of policies to maximize desired outcomes, such as throughput, security, or user satisfaction.

Consider a scenario where an RL agent manages access control and routing policies in a large data center. When faced with a DDoS attack, the agent dynamically reroutes traffic, adjusts firewall rules, and segments the network to contain the threat—all while learning from the effectiveness of its actions to improve future responses.

Deep Learning for Context-Aware Policy Enforcement

Deep learning models, with their ability to process high-dimensional data, are well-suited to complex policy scenarios involving diverse applications, devices, and user behaviors. These models can correlate information from multiple sources—such as IoT sensor data, user access logs, and application requirements—to enforce context-aware policies.

For instance, a deep neural network might detect that certain IoT devices are exhibiting unusual communication patterns, signaling a potential compromise. The SDN controller can then automatically segment these devices, apply stricter security policies, and alert administrators, all without human intervention.

Natural Language Processing for Intent-Based Policy Management

Natural language processing (NLP) bridges the gap between human operators and machine-executable policies. By allowing administrators to express desired outcomes in plain language, NLP models can parse these intents and translate them into precise SDN rules.

This capability reduces the risk of misconfiguration and accelerates policy deployment. An operator might simply state, "Prioritize telemedicine traffic during business hours," and the system automatically generates and enforces the corresponding QoS and routing policies.

Explainable AI for Trust and Compliance

As AI-driven policy management becomes more autonomous, the need for transparency and explainability increases. Explainable AI (XAI) techniques provide clear rationales for policy changes, helping operators and auditors understand why certain decisions were made. This is especially important in regulated industries where policy actions must be auditable and compliant with legal standards.

Table 7-1. *Major AI Techniques and Their Roles*

AI Technique	Role in Policy Management	Example Use Case
Machine Learning	Pattern recognition, impact prediction	Traffic-aware QoS adjustment
Reinforcement Learning	Adaptive policy optimization	Dynamic DDoS mitigation
Deep Learning	Context-aware, multidimensional policy control	IoT and user-segment-aware security policies
Natural Language Processing	Intent translation, automation	Human-to-policy interface
Explainable AI	Transparency, auditability	Regulatory compliance, operator trust

By combining these AI techniques, SDN environments gain the ability to move from static, rule-based management to a dynamic, self-optimizing paradigm that is responsive to both internal and external changes.

Case Studies of Dynamic Policy Management

The practical benefits of AI-powered dynamic policy management are already being realized across a variety of industries and network environments. The following case studies illustrate how organizations are leveraging these technologies to achieve greater agility, security, and efficiency.

Dynamic QoS Management in a Large University Campus

A prominent university faced persistent network congestion during peak periods, particularly when students streamed video lectures or conducted large-scale research data transfers. Traditional static QoS policies were unable to adapt quickly enough, resulting in degraded service quality and frequent complaints.

To address this, the university implemented an SDN solution integrated with machine learning-based QoS management. The system continuously monitored traffic patterns, predicted congestion based on historical and real-time data, and automatically adjusted bandwidth allocations to prioritize critical applications. As a result, user complaints dropped by 40%, remote learning experiences improved, and IT staff were able to focus on strategic initiatives rather than manual policy adjustments.

AI-Driven Security Policy Adaptation in a Financial Institution

A multinational financial institution operated in a highly regulated environment, facing constant threats from cyber attackers. Static security policies were insufficient to keep pace with evolving tactics, and manual updates often lagged behind the threat landscape.

The bank deployed an SDN platform enhanced with reinforcement learning and machine learning for security policy adaptation. The system detected anomalous access patterns, dynamically adjusted firewall and segmentation policies, and used explainable AI to provide compliance officers with clear justifications for each change. This approach resulted in a 70% reduction in successful phishing and lateral movement attacks and prevented data exfiltration by identifying anomalous outbound traffic patterns, automatic network segmentation, streamlined compliance audits, and a faster, more effective response to emerging threats.

Intent-Based Policy Management in a Global Enterprise WAN

A global logistics company managed a wide area network (WAN) spanning dozens of countries and serving diverse business units. The complexity of deploying and maintaining tailored policies for each unit led to delays and operational bottlenecks.

By adopting intent-based networking powered by NLP and deep learning, the company empowered business users to specify their requirements in natural language. The system translated these intents into SDN policies and optimized routing and resource allocation using deep learning models. Continuous monitoring ensured that policies adapted to shifting business priorities and network conditions. Policy deployment times dropped from days to minutes, and business units received customized network experiences without IT bottlenecks.

Multicloud Dynamic Policy Orchestration for a SaaS Provider

A leading SaaS provider needed to maintain consistent security and compliance policies across hybrid and multicloud environments. Manual policy management was slow, error-prone, and unable to keep up with the pace of cloud deployments.

The provider implemented AI-powered policy orchestration, aggregating telemetry from AWS, Azure, and on-premises SDN fabrics. Machine learning models identified policy conflicts and compliance gaps, while automated adjustments ensured a uniform security posture and regulatory compliance across all clouds. This led to a 50% reduction in policy-related security incidents, simplified audits, and a seamless user experience regardless of hosting location.

Automated Policy Conflict Resolution in a Healthcare Network

A large healthcare provider operated a complex network supporting electronic health records (EHR), telemedicine, and medical IoT devices. Policy conflicts between security, performance, and compliance requirements frequently led to service disruptions and increased operational risk.

By integrating deep learning and explainable AI into their SDN environment, the provider was able to automatically detect and resolve policy conflicts. The system analyzed the impact of conflicting rules, suggested optimal resolutions, and provided clear explanations for each action. This reduced service disruptions by 35%, improved compliance with healthcare regulations, and enhanced the overall reliability of critical medical applications.

Table: Summary of Case Study Outcomes

Table 7-2. *Quantitative Impact: Metrics and KPIs*

Organization/Use Case	AI Approach Used	Key Benefits Achieved
University Campus (QoS)	ML-based QoS management	Fewer complaints, improved learning/research
Financial Institution (Security)	RL, ML, XAI	Fewer breaches, compliance, faster response
Global Enterprise (Intent-Based WAN)	NLP, deep learning	Rapid deployment, tailored experiences
SaaS Provider (Multicloud)	ML, orchestration	Fewer incidents, compliance, seamless operations
Healthcare Network (Conflict Res.)	Deep learning, XAI	Fewer disruptions, compliance, reliability

The adoption of AI-powered dynamic policy management yields measurable improvements across key performance indicators.

Table 7-3. *KPI Improvements with AI-Driven Policy Management*

KPI	Before AI-Driven Policy	After AI-Driven Policy	Improvement
Policy Deployment Time	2–5 days	10–30 minutes	95% faster
Security Incident Rate	14/month	4/month	71% reduction
User Satisfaction (NPS)	6.5/10	8.9/10	+2.4 points
Operational Overhead	High	Moderate/low	Significant reduction
Compliance Audit Duration	3 weeks	1 week	67% faster

Note   NPS is an empirical metric used to measure customer loyalty and satisfaction. Organizations consistently report that AI-driven policy management not only boosts agility and security but also frees up human resources for innovation and strategic projects.

Lessons Learned and Best Practices

Successful deployments of AI-powered dynamic policy management reveal several best practices. Organizations should begin by identifying high-impact policy areas where dynamic adaptation will yield the greatest benefit, such as security or QoS. Ensuring the quality and timeliness of network telemetry is critical, as AI models rely on accurate data for effective decision-making.

Transparency is essential—operators and auditors must be able to trust and understand AI-driven policy changes, making explainable AI a key component. For critical or high-risk policy changes, a human-in-the-loop approach can balance automation with oversight. Continuous monitoring, analysis, and refinement of both policies and AI models ensure that the system evolves alongside changing business and technical requirements.

Conclusion

AI-powered dynamic policy management is redefining the capabilities of SDN, moving beyond static, manual processes to deliver networks that are agile, secure, and resilient. By integrating machine learning, reinforcement learning, deep learning, NLP, and explainable AI, organizations can create policy frameworks that sense, predict, and respond to change in real time.

This chapter has laid out the principles of policy management in SDN, explored the AI techniques enabling dynamic adaptation, and illustrated real-world successes. As networks continue to evolve, embracing AI-driven policy management will be essential for meeting the demands of digital business, defending against sophisticated threats, and unlocking new levels of efficiency and user satisfaction. The journey towards fully autonomous, self-optimizing networks is well underway, and those who lead the way will gain a decisive advantage in the digital age.

CHAPTER 8

Reinforcement Learning for Autonomous Traffic Engineering

The exponential growth in Internet-connected devices, cloud services, and multimedia applications has placed unprecedented demands on network infrastructure. Traditional traffic engineering approaches, which rely on static rules and manual configuration, often struggle to adapt to such dynamic and complex environments. The need for networks that can intelligently and autonomously manage the flow of data has never been greater.

Reinforcement learning (RL), a branch of artificial intelligence, offers a compelling solution. By enabling systems to learn optimal behaviors through interaction with their environment, RL empowers networks to make real-time decisions, adapt to changing conditions, and optimize performance without human intervention. This chapter delves into the principles of reinforcement learning, explores its application to traffic engineering, presents real-world case studies, and considers future directions for autonomous network management.

Overview of Reinforcement Learning

Reinforcement learning is a paradigm of machine learning where an agent learns to make decisions by performing actions and receiving feedback in the form of rewards or penalties. Unlike supervised learning, which requires labeled datasets, RL thrives in scenarios where the correct actions are not known in advance and must be discovered through exploration.

© Het Mehta 2026
H. Mehta, *AI Agents for Secure and Software-Defined Networking,*
https://doi.org/10.1007/979-8-8688-2358-9_8

At the heart of RL are several fundamental components:

- **Agent:** The decision-maker (e.g., an SDN controller)

- **Environment:** The system within which the agent operates (e.g., a network)

- **State:** The current situation of the environment (e.g., network topology, traffic load)

- **Action:** Choices the agent can make (e.g., reroute flows, adjust bandwidth)

- **Reward:** Feedback signal indicating the desirability of an action (e.g., reduced latency, increased throughput)

- **Policy:** The strategy used by the agent to select actions based on states

The process is iterative. The agent observes the state of the environment, selects an action, receives a reward, and updates its policy to improve future decisions. This iterative cycle allows the agent to gradually learn the most effective strategies for achieving its goals, even in environments that are constantly changing or partially observable.

Reinforcement Learning Loop

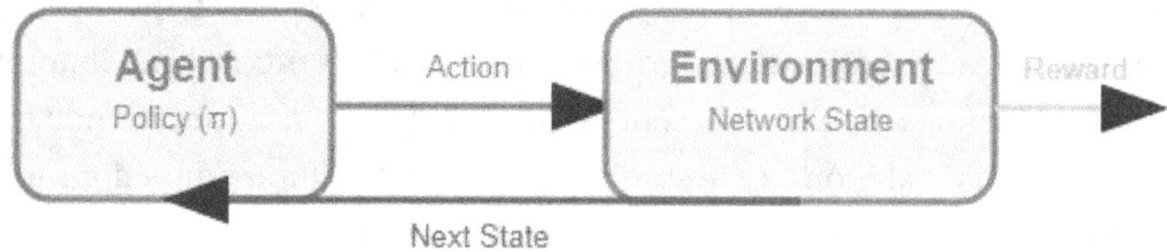

Figure 8-1. *The Reinforcement Learning Loop*

The RL agent observes the environment, takes an action, receives a reward, and updates its policy based on the outcome, forming a continuous feedback cycle.

The RL agent's objective is to maximize cumulative rewards over time. This is typically formalized as a Markov Decision Process (MDP), where the next state and reward depend only on the current state and action. The iterative nature of RL enables agents to improve their strategies through trial and error. Over time, the agent learns which actions yield higher rewards in specific situations, leading to increasingly effective behavior.

Markov Decision Process in RL

Figure 8-2. *Markov Decision Process in RL*

An RL agent transitions from one state to another by taking actions and receiving rewards, forming the foundation of its learning process.

Reinforcement learning encompasses a variety of algorithms, each with unique strengths. The choice of algorithm depends on the complexity of the environment, the available computational resources, and the specific requirements of the traffic engineering task.

Table 8-1. *Types of Reinforcement Learning Algorithms*

Algorithm Type	Description	Example Use Case
Value-Based (e.g., Q-Learning)	Learns the value of actions in each state	Routing in dynamic topologies
Policy-Based	Directly learns the policy for action selection	Resource allocation
Actor-Critic	Combines value and policy methods for efficiency	Congestion control
Deep RL	Uses neural networks to handle large state spaces	Complex, high-dimensional networks

This table summarizes the main categories of RL algorithms and their typical applications in networking.

Traffic Engineering Strategies

Traffic engineering (TE) refers to the process of optimizing the flow of data across a network to achieve desired performance objectives such as minimizing latency, maximizing throughput, or ensuring fairness among users. Traditionally, TE relied on static rules, manual configuration, and periodic adjustments based on historical data. However, these methods struggle to cope with today's volatile traffic patterns and diverse application demands.

The integration of RL into traffic engineering enables networks to autonomously sense changes, predict outcomes, and reconfigure themselves in real time. RL-based TE can outperform static methods, especially in dynamic environments.

The contrast between traditional and RL-based TE is striking.

Table 8-2. *Comparison of Traditional and RL-Based Traffic Engineering*

Aspect	Traditional TE	RL-Based TE
Adaptability	Low (manual/static)	High (real-time, autonomous)
Scalability	Challenging at large scale	Handles large, complex networks
Responsiveness	Slow (human-in-the-loop)	Fast (machine-driven)
Optimization	Rule-based, often suboptimal	Learns optimal strategies
Data Utilization	Historical, limited	Real-time, continuous

Traditional TE relies on static, manual decisions, while RL-based TE leverages AI for adaptive, autonomous optimization.

Traditional vs. RL-Based Traffic Engineering

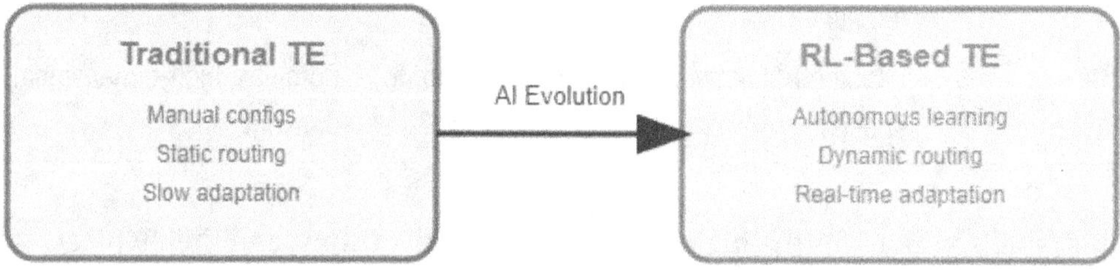

Figure 8-3. *Traditional vs. RL-Based Traffic Engineering*

This figure visually contrasts the static, manual nature of traditional TE with the adaptive, automated approach of RL-based TE.

An RL-driven TE system typically operates as follows:

- **State Observation:** The agent collects real-time metrics such as link utilization, queue lengths, and flow statistics.

- **Action Selection:** Based on its policy, the agent chooses actions like rerouting flows, adjusting priorities, or allocating bandwidth.

- **Reward Evaluation:** After actions are taken, the agent receives feedback (e.g., lower latency, fewer dropped packets) and updates its policy.

- **Policy Update:** The RL model is refined to improve future decisions, creating a feedback loop that drives continuous optimization.

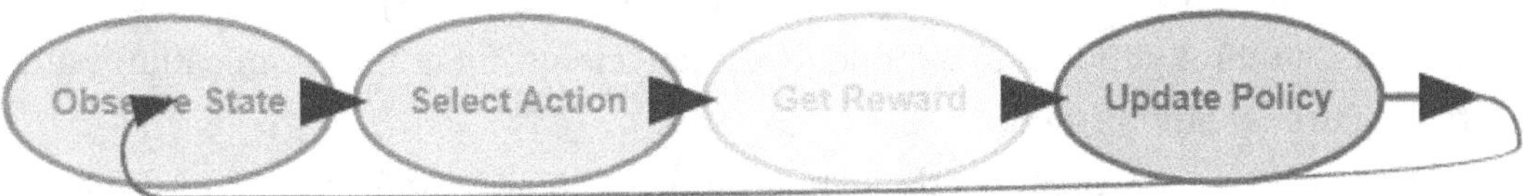

Figure 8-4. *RL-Driven TE Workflow*

The RL agent observes the network state, selects an action, receives a reward, and updates its policy in a continuous, adaptive loop.

RL has been successfully applied to several traffic engineering challenges.

Table 8-3. *RL Applications in Traffic Engineering*

Application	RL Role	Key Benefits
Dynamic Routing	Selects the best paths in real time	Lower latency, higher throughput
Congestion Control	Adjusts rates based on feedback	Fewer dropped packets, stability
Load Balancing	Distributes flows to prevent hot spots	Better resource utilization
QoS Enforcement	Allocates bandwidth per application	Improved user experience

This table summarizes common RL applications in TE and their benefits.

Case Studies in Autonomous Traffic Management

The application of RL to traffic engineering is moving rapidly from research labs to operational networks. The following case studies demonstrate the transformative impact of RL-driven solutions.

A leading Internet Service Provider deployed an RL agent within its SDN-based backbone to optimize inter-city traffic flows. The agent continuously monitored link loads, failure events, and latency metrics. It learned to reroute traffic dynamically, minimizing congestion during peak hours and rapidly responding to outages. The result was an 18% reduction in average end-to-end latency, a 25% decrease in packet loss during peak periods, and faster recovery from network failures.

A global cloud provider implemented RL-driven congestion control in its data center networks. The RL agent managed queue lengths and adjusted transmission rates based on real-time telemetry, learning to preemptively avoid hotspots. This led to a 30% increase in throughput, a 40% reduction in tail latency for critical applications, and improved fairness among tenants.

A large-scale video streaming platform used RL to balance traffic across multiple content delivery paths. The agent learned to anticipate surges in demand and proactively shifted flows, ensuring consistent quality for viewers. The system saw 22% fewer buffering events, a 15% increase in user satisfaction scores, and more efficient utilization of network resources.

RL in Real-World Traffic Engineering

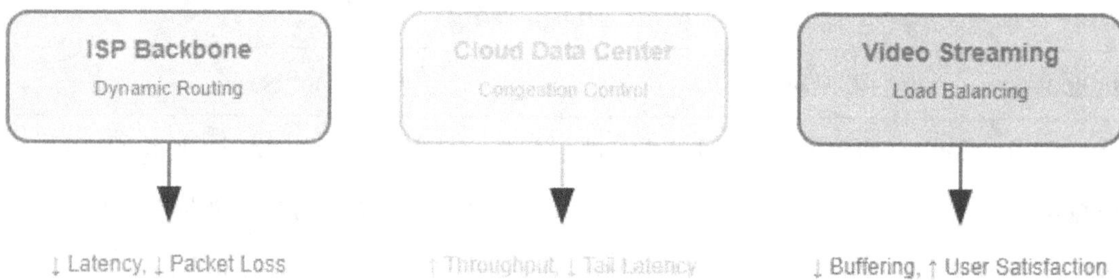

Figure 8-5. *RL in Real-World Traffic Engineering*

Examples of RL-driven solutions in ISP backbones, cloud data centers, and video streaming networks, each delivering measurable improvements.

Table 8-4. *Summary of RL-Driven Traffic Engineering Case Studies*

Organization/Use Case	RL Application	Key Results
Tier-1 ISP Backbone	Dynamic Routing	Lower latency, less packet loss
Cloud Data Center	Congestion Control	Higher throughput, lower latency
Video Streaming Platform	Load Balancing	Fewer buffering events, happier users

This table highlights the diverse impact RL can have on real-world network performance.

Future Directions in Traffic Engineering

As reinforcement learning continues to mature, its role in autonomous traffic engineering is set to expand. Several emerging trends and research areas are shaping the future of this field.

While many current systems use a single RL agent, future networks will likely deploy multiple agents working collaboratively or competitively. Multi-agent RL can enable decentralized decision-making, enhance scalability, and improve fault tolerance. For example, each network domain could have its own agent, with agents coordinating to achieve global objectives.

Combining RL with other AI methodologies, such as supervised learning or unsupervised anomaly detection, can create more robust and versatile traffic engineering solutions. Hybrid models may leverage historical data for initial training and then use RL for ongoing adaptation.

As RL agents become more autonomous, understanding and trusting their decisions becomes critical. Research into explainable RL aims to provide network operators with insights into why certain actions are taken, facilitating troubleshooting and compliance.

Future RL systems will need to handle increasingly heterogeneous environments, including hybrid cloud, IoT, and 5G/6G networks. Real-time adaptation across diverse technologies and administrative domains will be essential for end-to-end optimization.

Future Trends in RL-Driven TE

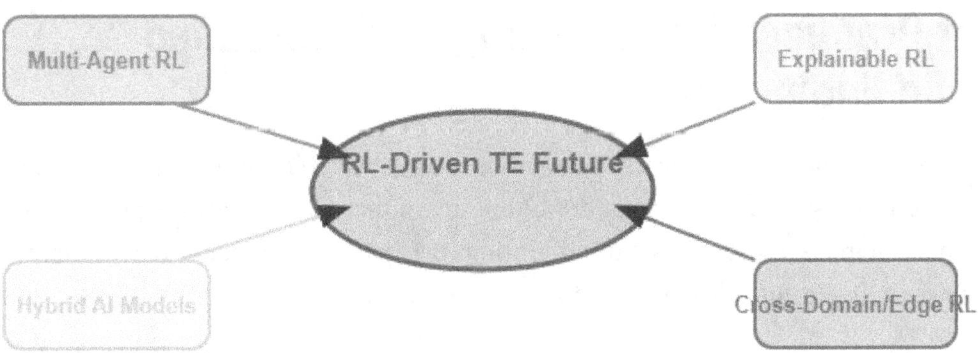

Figure 8-6. *Future Trends in RL-Driven Traffic Engineering*

Emerging directions include multi-agent RL, hybrid AI models, explainable RL, and cross-domain/edge-centric RL strategies.

Table 8-5. *Emerging Trends in RL for Traffic Engineering*

Trend	Description	Potential Impact
Multi-agent RL	Multiple agents coordinate decisions	Scalability, resilience
Hybrid AI Models	Combine RL with other AI approaches	Improved accuracy, flexibility
Explainable RL	Transparent, auditable decision-making	Trust, compliance
Cross-Domain Adaptation	Operate across varied network types	End-to-end performance
Edge-Centric RL	RL at the network edge for low-latency adaptation	Real-time, localized control

This table summarizes future research directions and their expected impact on network autonomy.

Reinforcement learning is revolutionizing the field of traffic engineering by enabling autonomous, adaptive, and intelligent network management. By learning from experience and optimizing decisions in real time, RL agents are delivering significant improvements in performance, reliability, and user satisfaction. As networks become more complex and dynamic, the integration of RL with other AI techniques and the development of explainable, multi-agent systems will drive the next wave of innovation in autonomous traffic engineering.

Best Practices and Challenges in RL-Based Traffic Engineering

As RL-based traffic engineering systems move from research to deployment, network architects and engineers must consider both the unique strengths and the practical challenges of these approaches. Implementing RL in production networks requires careful planning, robust design, and ongoing monitoring.

Best Practices

- **Start with Simulation:** Before deploying RL agents in live networks, thoroughly test them in simulated environments that mimic real traffic patterns and failure scenarios. This reduces risks and allows fine-tuning of reward functions and policies.

- **Define Clear Reward Metrics:** The choice of the reward function directly influences agent behavior. Combine multiple objectives— such as latency, throughput, and fairness—using weighted sums or multi-objective optimization.

- **Monitor and Audit Decisions:** Implement logging and monitoring to track RL agent decisions. This enables operators to audit unexpected actions and refine the agent's learning process.

- **Incremental Deployment:** Introduce RL agents gradually, starting with noncritical paths or as advisory systems. Gradually expand their scope as confidence grows.

- **Human-in-the-Loop:** In early stages, keep operators involved in reviewing and approving agent actions, especially for high-impact decisions. This hybrid approach ensures safety and builds trust.

Common Challenges

While RL offers tremendous promise, several challenges must be addressed for successful adoption:

- **Exploration vs. Exploitation:** RL agents must balance exploring new strategies with exploiting known good actions. Too much exploration may disrupt service, while too little can prevent finding optimal solutions.

- **Scalability:** As network size and complexity grow, the state and action spaces can become enormous. Deep RL and hierarchical approaches can help, but require significant computational resources.

- **Delayed Rewards:** In networks, the impact of an action (like rerouting traffic) may not be immediately visible. Designing reward structures that account for delayed effects is nontrivial.

- **Safety and Stability:** RL agents may occasionally make risky or suboptimal decisions, especially early in training. Safeguards and fallback mechanisms are essential to prevent network outages.

- **Explainability:** Operators often require explanations for agent actions, especially when outcomes are unexpected. Integrating explainable AI techniques is an active area of research.

Table 8-6. *Key Challenges and Mitigation in RL-Based Traffic Engineering*

Challenge	Description	Mitigation Strategies
Exploration/ Exploitation	Balancing learning new vs. using known actions	Use safe exploration, epsilon-greedy policies
Scalability	Large state/action spaces in big networks	Deep RL, hierarchical RL, state abstraction
Delayed Rewards	Actions' effects may be seen much later	Shaped rewards, eligibility traces
Safety	Avoiding risky or disruptive decisions	Human oversight, constraints, rollback plans
Explainability	Making the agent's decisions understandable	Logging, visualization, XAI methods

These best practices and mitigation strategies help ensure that RL deployments are robust, trustworthy, and aligned with business and operational goals.

Transitioning to Autonomous Networks

The deployment of RL for traffic engineering is often a stepping stone toward fully autonomous, self-driving networks. In such networks, AI agents not only optimize traffic flows but also handle configuration, security, fault management, and resource provisioning with minimal human intervention.

A gradual transition is recommended:

- **Phase 1:** Use RL for advisory roles, providing recommendations to human operators.

- **Phase 2:** Allow RL agents to manage selected segments or services under supervision.

- **Phase 3:** Expand RL autonomy to core network operations, with humans monitoring and intervening only for exceptions.

This staged approach helps organizations build confidence in AI-driven automation, refine operational processes, and address regulatory or compliance concerns.

Looking ahead, RL will increasingly be integrated with other AI techniques—such as supervised learning for traffic prediction, unsupervised learning for anomaly detection, and natural language processing for intent-driven networking. The result will be networks that not only react to current conditions but also anticipate future needs and proactively optimize themselves.

Conclusion

Reinforcement learning is rapidly becoming a cornerstone technology for autonomous traffic engineering, enabling networks to optimize performance dynamically and intelligently. By continuously learning from network states and adapting decisions in real time, RL-driven systems improve latency, throughput, and resilience beyond what traditional methods can achieve. The case studies presented demonstrate practical successes and highlight the transformative potential of RL in diverse network environments.

As networks grow increasingly complex and heterogeneous, future developments will focus on multi-agent collaboration, hybrid AI models, explainability, and cross-domain adaptability. These advancements will further empower networks to self-manage, self-heal, and proactively address emerging challenges.

Building on the foundations of RL for traffic engineering, the next chapter explores **ethical AI agents for responsible network management**, emphasizing the importance of governance, transparency, and trust in deploying AI-driven network solutions.

CHAPTER 9

Ethical AI Agents for Responsible Network Management

The integration of artificial intelligence (AI) agents into network management has revolutionized how networks are monitored, optimized, and secured. These AI-driven systems can analyze complex data, make autonomous decisions, and adapt dynamically to changing conditions. However, with this increased autonomy comes a profound responsibility to ensure that AI operates ethically and transparently. Ethical AI deployment is essential to safeguard user rights, maintain trust, and comply with regulatory requirements.

This chapter explores the ethical landscape surrounding AI agents in network management. We begin by establishing the core ethical principles relevant to AI deployment and identifying the challenges unique to networking environments. Then, we examine prominent ethical AI frameworks that provide structured guidance for responsible AI development and deployment. Finally, we discuss governance models and compliance strategies that organizations can adopt to ensure ethical and lawful AI operations.

By embedding ethics into AI-driven network management, organizations can build resilient, trustworthy, and fair networks that serve both business objectives and societal values.

© Het Mehta 2026
H. Mehta, *AI Agents for Secure and Software-Defined Networking,*
https://doi.org/10.1007/979-8-8688-2358-9_9

Ethics in AI Deployment

Ethics forms the backbone of responsible AI deployment. Without a clear ethical foundation, AI agents risk making decisions that could harm users, violate privacy, or introduce unfairness in network operations. Understanding the fundamental ethical principles and challenges is the first step toward designing AI systems that align with societal values and legal requirements.

Core Ethical Principles for AI in Networking

Ethical AI deployment is grounded in universal principles that guide the behavior of AI agents and the organizations that develop and deploy them. The following principles are particularly important in the context of network management:

- **Fairness:** AI agents must treat all users, devices, and applications equitably. This means avoiding biases that could lead to unfair prioritization or discrimination, such as favoring certain traffic types or geographic regions unfairly.

- **Transparency:** Network operators and stakeholders should be able to understand and interpret AI decisions. Transparency fosters trust and enables effective troubleshooting and accountability.

- **Privacy:** Protecting user data is critical, especially given the sensitive nature of network traffic information. AI systems must adhere to privacy laws and use techniques like data anonymization and encryption.

- **Accountability:** Clear lines of responsibility must be established for AI actions. When AI decisions lead to negative outcomes, organizations must be able to investigate, explain, and rectify issues.

- **Safety and Security:** AI agents should operate reliably without causing unintended disruptions or opening vulnerabilities that could be exploited by attackers.

- **Beneficence:** AI should actively contribute to improving network performance, reliability, and user experience while minimizing harm.

These principles create a framework for ethical decision-making and system design, ensuring AI advances network management without compromising ethical standards.

Unique Ethical Challenges in Network AI

While the core principles provide a foundation, the practical deployment of AI in networking introduces specific ethical challenges. These challenges arise from the complexity of networks, the sensitivity of data involved, and the critical nature of network services.

Some of the most pressing ethical challenges include algorithmic bias, opaque decision-making processes, privacy concerns, balancing AI autonomy with human oversight, and safeguarding against security vulnerabilities introduced by AI itself. Understanding these challenges helps in designing AI agents that are not only effective but also ethically responsible.

Embedding Ethical Decision-Making in AI Agents

To translate ethical principles into actionable AI behavior, ethical decision-making must be embedded within AI agents. This involves aligning AI objectives with human values, incorporating reasoning mechanisms to evaluate ethical trade-offs, and ensuring human oversight remains part of the decision process.

Embedding ethics into AI agents helps prevent unintended consequences and builds trust among network operators and users. It also prepares AI systems to adapt to evolving ethical standards and societal expectations over time.

Frameworks for Ethical AI in Networking

Navigating the complex ethical landscape requires structured guidance. Ethical AI frameworks provide this guidance by outlining principles, best practices, and compliance requirements. These frameworks help organizations design, develop, and deploy AI systems responsibly, ensuring alignment with legal and societal norms.

Overview of Leading Ethical AI Frameworks

Several internationally recognized frameworks offer comprehensive guidelines for ethical AI. These frameworks emphasize transparency, fairness, privacy, and accountability and can be adapted to the specific needs of network AI. Understanding these frameworks is crucial for organizations aiming to implement ethical AI governance.

Tailoring Frameworks to Network AI Agents

Applying general ethical AI frameworks to the specific context of networking involves adapting principles to address the unique challenges of network management. For instance, transparency must be operationalized through explainable AI tools tailored to network decisions, and privacy must be strictly enforced given the sensitivity of network data.

Tailoring frameworks ensures that ethical AI principles are not just theoretical but are practically integrated into network AI agents' design and operation.

Ethical Auditing, Certification, and Continuous Compliance

Maintaining ethical standards is an ongoing process. Ethical auditing, certification, and continuous monitoring provide mechanisms to ensure AI systems remain compliant with evolving ethical and regulatory expectations. These processes help organizations identify risks early, demonstrate accountability, and build stakeholder confidence.

Governance and Compliance in AI

Effective governance and compliance frameworks are essential to operationalize ethical AI principles. They establish clear roles, policies, and procedures that guide the responsible development and use of AI agents in network management.

Governance Models for AI in Networking

Governance models define how organizations oversee AI initiatives. Whether centralized, decentralized, or hybrid, governance structures ensure that ethical considerations are integrated into every stage of the AI lifecycle management. Choosing an appropriate governance model depends on organizational context and regulatory requirements.

Regulatory Compliance

AI systems must operate within the boundaries set by laws and regulations. This includes data protection laws like GDPR and CCPA, telecommunications regulations, and emerging AI-specific legislation. Compliance requires embedding legal requirements into AI system design and organizational processes.

Organizational Policies and Best Practices

Beyond legal compliance, organizations should develop internal policies that promote ethical AI use. Training, incident response protocols, and stakeholder engagement are key components that embed ethics into organizational culture and daily operations.

Tools and Technologies Supporting Governance

Technological advancements such as explainable AI, automated compliance tools, audit trails, and simulation platforms support governance efforts by enhancing transparency, traceability, and control over AI systems.

Ethical AI Frameworks Overview

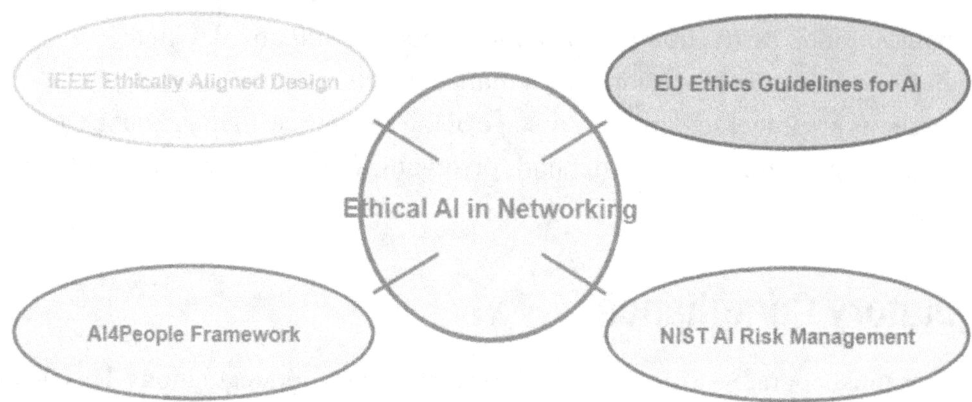

Figure 9-1. *Ethical AI Frameworks Overview*

This diagram illustrates the major ethical AI frameworks that provide foundational guidance for the responsible design and deployment of AI agents in networking environments.

Conclusion

Ethical AI deployment is a critical pillar for responsible network management. By adhering to core principles such as fairness, transparency, privacy, and accountability, AI agents can enhance network performance while safeguarding user rights and organizational integrity. Established ethical frameworks provide valuable guidance, but their successful application requires careful adaptation to the unique challenges of networking.

Robust governance models and compliance mechanisms ensure AI systems operate lawfully and ethically, fostering trust among stakeholders. As AI-driven network management continues to evolve, embedding ethics at every stage—from design to deployment and monitoring—will be essential to harness AI's full potential responsibly.

The next chapter will explore **AI agents for network threat intelligence and response**, demonstrating how ethical AI principles underpin security applications that protect networks from increasingly sophisticated cyber threats.

CHAPTER 10

AI Agents for Network Threat Intelligence and Response

In today's hyperconnected digital landscape, network security faces unprecedented challenges. Cyber adversaries employ increasingly sophisticated tactics, techniques, and procedures (TTPs) that evolve rapidly, making traditional security approaches insufficient. The volume of security data generated by modern networks is immense, overwhelming human analysts and manual processes. Artificial intelligence (AI) agents have thus become essential allies in the ongoing battle to protect critical infrastructure, sensitive data, and user privacy.

AI agents bring remarkable capabilities to network threat intelligence and response by automating data collection, analyzing complex patterns, detecting anomalies, and orchestrating timely responses. Their ability to learn from vast datasets, adapt to new threats, and collaborate across organizational boundaries transforms how security teams anticipate, identify, and mitigate cyber risks.

This chapter delves into how AI agents empower network threat intelligence and response. We start by outlining foundational threat intelligence concepts, followed by a detailed survey of AI techniques for threat detection. Then, we explore incident response strategies enhanced by AI, emphasizing automation and prioritization. Collaboration mechanisms that leverage AI to share intelligence securely are discussed next. Finally, we examine AI's critical role in certificate and key management, vital for securing network communications.

By understanding and leveraging these AI-driven approaches, organizations can build resilient defenses that keep pace with evolving cyber threats while maintaining ethical and operational integrity.

© Het Mehta 2026
H. Mehta, *AI Agents for Secure and Software-Defined Networking*,
https://doi.org/10.1007/979-8-8688-2358-9_10

Threat Intelligence Concepts

Threat intelligence is the cornerstone of proactive cybersecurity. It transforms raw data into actionable knowledge about adversaries, their methods, and potential targets, enabling organizations to anticipate and thwart attacks before they cause damage.

Types of Threat Intelligence

Threat intelligence is commonly divided into three interrelated categories, each serving distinct but complementary purposes:

- **Strategic Intelligence:** This high-level intelligence focuses on the broader threat landscape, including the motivations, capabilities, and objectives of threat actors. It informs leadership and security architects to shape policies, investments, and long-term defense strategies. For example, understanding nation-state cyber espionage trends helps governments allocate resources effectively.

- **Tactical Intelligence:** Tactical intelligence provides detailed insights into the tactics, techniques, and procedures (TTPs) used by attackers. This intelligence supports the development of detection rules, signatures, and security controls. For instance, knowing that a ransomware group exploits a specific vulnerability enables rapid patching and monitoring.

- **Operational Intelligence:** This real-time intelligence relates to active threats and ongoing campaigns targeting specific assets or sectors. It supports immediate response actions and incident handling. For example, alerts about a distributed denial of service (DDoS) attack targeting a financial institution's website trigger mitigation protocols.

AI agents can ingest and correlate information across these intelligence types, providing a holistic and timely picture of the threat environment.

Sources of Threat Intelligence

Effective threat intelligence relies on diverse, credible data sources:

- **Internal Network Logs and Telemetry:** Network devices, firewalls, intrusion detection/prevention systems (IDS/IPS), endpoint security tools, and application logs provide rich data about network activity and anomalies. AI agents analyze these sources to detect suspicious behaviors.

- **External Threat Feeds:** Commercial and open source feeds supply indicators of compromise (IOCs), malware hashes, phishing URLs, and vulnerability disclosures. Integrating these feeds enhances situational awareness and detection capabilities.

- **Dark Web and Underground Forums:** Cybercriminals often communicate and trade exploits on dark web marketplaces and forums. AI-powered natural language processing (NLP) tools mine these sources to extract emerging threats and attacker chatter.

- **Open Source Intelligence (OSINT):** Publicly available information from social media, blogs, security advisories, and research reports enriches threat context and attribution efforts.

AI agents excel at aggregating, normalizing, and analyzing this heterogeneous data, enabling timely detection of threats that might otherwise go unnoticed.

The Intelligence Cycle

The threat intelligence lifecycle is a continuous process that transforms raw data into actionable insights:

- **Collection:** Automated AI agents gather data from multiple sources, ensuring comprehensive coverage.

- **Processing:** Data is cleansed, normalized, and enriched to facilitate analysis.

- **Analysis:** AI models correlate data points, identify patterns, and generate intelligence reports or alerts.

- **Dissemination:** Actionable intelligence is shared with security teams, automated defense systems, or trusted partners.

- **Feedback:** Insights from incident outcomes and analyst input refine AI models, improving future intelligence quality.

AI accelerates and automates each phase, enabling continuous, adaptive threat intelligence that keeps pace with attacker innovation.

AI Techniques for Threat Detection

Detecting threats in complex network environments requires AI techniques capable of processing vast data volumes, recognizing subtle patterns, and adapting to novel attack vectors.

Machine Learning for Anomaly Detection

Machine learning (ML) is foundational for identifying deviations from normal network behavior, which often signal malicious activity.

- **Supervised Learning:** Models are trained on labeled data sets containing examples of benign and malicious traffic. Algorithms such as decision trees, support vector machines (SVMs), and random forests classify traffic based on learned features. While effective for known threats, supervised learning depends heavily on quality labeled data.

- **Unsupervised Learning:** Without labeled data, algorithms like clustering (k-means, DBSCAN) and autoencoders detect anomalies by identifying outliers or patterns that differ from established baselines. This approach is valuable for discovering zero-day attacks or insider threats.

- **Semi-supervised Learning:** Combines both approaches, using limited labeled data alongside large unlabeled datasets to improve detection accuracy when labeled data is scarce.

ML models continuously refine their understanding of "normal" network behavior, enabling dynamic anomaly detection.

Deep Learning for Pattern Recognition

Deep learning extends ML capabilities by automatically extracting hierarchical features from raw data, excelling at complex pattern recognition.

- **Convolutional Neural Networks (CNNs):** Originally designed for image processing, CNNs are adapted to analyze network traffic payloads, identifying malware signatures embedded in packet data.

- **Recurrent Neural Networks (RNNs) and Long Short-Term Memory (LSTM):** These architectures model sequential data, capturing temporal dependencies in event logs or network flows. They detect multistage attacks that unfold over time, such as advanced persistent threats (APTs).

Deep learning models require significant computational resources and large datasets but deliver high detection accuracy, especially for sophisticated threats.

Natural Language Processing (NLP) for Threat Intelligence

NLP techniques enable AI agents to extract actionable insights from unstructured textual data sources:

- Parsing security advisories, vulnerability reports, and patch notes to identify relevant threats

- Monitoring dark web forums and social media for emerging attack campaigns and threat actor communications

- Automating the classification and summarization of threat intelligence reports to accelerate analyst workflows

By leveraging NLP, AI agents expand the scope of threat intelligence beyond structured network data.

Reinforcement Learning for Adaptive Defense

Reinforcement learning (RL) agents learn optimal defense strategies by interacting with simulated or real network environments:

- RL models receive feedback based on the success or failure of defense actions, such as blocking suspicious IPs or adjusting firewall rules.

- Over time, agents optimize their policies to minimize risk and maximize network availability.

RL offers promising potential for dynamic, self-improving security systems that adapt to changing threat landscapes.

Table 10-1. *AI Techniques and Their Applications in Threat Detection*

AI Technique	Description	Application Example
Supervised Learning	Uses labeled data to classify threats	Detecting known malware signatures
Unsupervised Learning	Identifies anomalies without labels	Discovering novel intrusion behaviors
Deep Learning (CNN, RNN)	Processes complex data and sequences	Analyzing traffic payloads for malware
NLP	Extracts insights from unstructured text	Parsing threat reports and dark web chatter
Reinforcement Learning	Learns adaptive strategies through trial and error	Dynamic firewall rule adjustment

Incident Response Strategies

Detection alone is insufficient; rapid and effective incident response minimizes damage and restores normal operations.

Automated Response Systems

AI agents automate routine and repetitive response tasks, reducing human workload and response times:

- Isolating compromised endpoints by severing network access

- Blocking malicious IP addresses or domains at firewalls

- Quarantining suspicious files detected on endpoints

- Initiating forensic data collection for further investigation

Automation enables near real-time containment of threats, limiting lateral movement and data loss.

Prioritization and Triage

Security teams face alert fatigue due to high volumes of security events. AI-driven incident prioritization helps focus efforts on the most critical threats:

- Risk scoring models assess the severity and confidence of alerts based on contextual data.

- Correlation engines group related alerts into incidents, reducing noise.

- Prioritized incident queues optimize analyst workflows and resource allocation.

This triage process improves operational efficiency and reduces the risk of missing high-impact threats.

Root Cause Analysis

Understanding how an attack occurred is vital for remediation and prevention:

1. AI agents analyze event sequences, network flows, and system logs to reconstruct attack chains.

2. Visualization tools highlight compromised assets and attack paths.

3. Root cause insights inform patching, configuration changes, and policy updates.

Accurate root cause analysis accelerates recovery and strengthens defenses.

Post-incident Learning

Each incident provides valuable data to improve future detection and response:

- AI models retrain using incident data to recognize similar threats.

- Lessons learned feed into threat intelligence sharing and training programs.

- Continuous improvement cycles enhance organizational resilience.

Collaboration in Threat Intelligence

No organization operates in isolation; sharing threat intelligence enhances collective security.

Information Sharing Platforms

Information Sharing and Analysis Centers (ISACs), government initiatives, and industry consortia provide trusted platforms for exchanging threat data:

- Members share IOCs, attack patterns, and mitigation strategies.

- Collaborative alerts accelerate detection across sectors.

- Shared intelligence helps identify emerging threats early.

AI agents facilitate ingestion, analysis, and dissemination of shared intelligence at scale.

Federated Learning for Privacy-Preserving Collaboration

Federated learning enables multiple organizations to collaboratively train AI models without exposing sensitive raw data:

- Each participant trains a local model on their data.

- Updates are aggregated centrally to build a global model.

- Raw data never leaves the local environment, preserving privacy.

This approach balances collaboration benefits with data protection requirements.

Standardization and Interoperability

Standard formats and protocols ensure seamless exchange and integration of threat intelligence:

- **STIX (Structured Threat Information eXpression):** A standardized language for representing threat information

- **TAXII (Trusted Automated eXchange of Indicator Information):** A protocol for exchanging threat intelligence data

Adopting standards enables AI systems to consume and share intelligence consistently and efficiently.

Collaboration in Threat Intelligence

Figure 10-1. *Collaboration in Threat Intelligence*

Multiple organizations share threat data via a central intelligence cloud, enabling collective defense and improved situational awareness.

AI for Certificate and Key Management

Cryptographic certificates and keys are fundamental to securing network communications, authenticating devices, and encrypting data. However, managing these assets at scale is complex and error-prone.

Challenges in Certificate and Key Management

- **Expiration and Renewal:** Certificates have limited validity periods. Failure to renew on time can cause service outages or security lapses.

- **Revocation and Compromise:** Keys may be compromised or misused, necessitating timely revocation to prevent unauthorized access.

- **Inventory Complexity:** Large enterprises may have thousands of certificates spread across devices, applications, and cloud services, complicating tracking.

- **Policy Compliance:** Regulatory frameworks require strict controls and audits over cryptographic assets.

Manual management is inefficient and vulnerable to human error, increasing operational risk.

AI-Driven Solutions

AI agents augment certificate and key management through

- **Automated Discovery and Inventory:** AI scans networks and cloud environments to detect all certificates and keys, including shadow IT assets unknown to administrators.

- **Predictive Renewal Scheduling:** Machine learning models forecast optimal renewal windows based on usage patterns and certificate lifetimes, preventing unexpected expirations.

- **Anomaly Detection:** AI monitors key usage for unusual patterns that may indicate compromise or misuse, triggering alerts and automated responses.

- **Policy Enforcement and Compliance:** AI verifies that certificates meet organizational policies, such as key lengths, algorithms, and trusted authorities, ensuring regulatory compliance.

By automating these functions, AI reduces risk, operational overhead, and enhances the security posture of cryptographic infrastructure.

Conclusion

Artificial intelligence agents have become indispensable in the modern cybersecurity arsenal, particularly in network threat intelligence and response. By harnessing advanced AI techniques such as machine learning, deep learning, natural language processing, and reinforcement learning, organizations can achieve rapid, accurate, and adaptive threat detection. AI-driven automation accelerates incident response, enabling containment and remediation that minimize damage.

Collaboration among organizations, empowered by AI technologies like federated learning and standardized intelligence sharing, strengthens collective defense against increasingly sophisticated cyber adversaries. Furthermore, AI's role in automating certificate and key management addresses critical security needs foundational to trusted network communications.

As the cyber threat landscape continues to evolve, integrating AI agents into network security strategies is essential for maintaining resilient, ethical, and effective defenses. The continued advancement of AI will play a pivotal role in safeguarding digital infrastructure and ensuring secure, reliable network operations.

CHAPTER 11

AI-Driven Resource Allocation in Multi-tenant SDN Environments

Software-Defined Networking (SDN) has emerged as a transformative paradigm in network architecture, enabling programmable, centralized control of network resources. By decoupling the control plane from the data plane, SDN allows network administrators to dynamically configure, manage, and optimize network behavior through software applications. This flexibility is particularly valuable in multi-tenant environments, where multiple independent tenants share a common physical infrastructure but require logical isolation and customized network policies.

Multi-tenant SDN environments are prevalent in cloud data centers, telecom networks supporting network slicing, and enterprise networks that host multiple business units or customers on the same infrastructure. These environments introduce complex resource allocation challenges as diverse tenant demands must be met with limited physical resources, all while maintaining strict isolation, performance guarantees, and security.

Traditional resource allocation methods based on static rules or heuristic algorithms struggle to cope with the scale, dynamism, and unpredictability inherent in multi-tenant SDNs. Artificial intelligence (AI), with its capabilities in learning, prediction, and optimization, offers powerful solutions to these challenges. AI-driven resource allocation can dynamically adapt to changing network conditions, forecast future demands, and optimize multiple conflicting objectives such as throughput, latency, fairness, and security.

© Het Mehta 2026
H. Mehta, *AI Agents for Secure and Software-Defined Networking*,
https://doi.org/10.1007/979-8-8688-2358-9_11

This chapter explores the multifaceted challenges of resource allocation in multi-tenant SDN environments, surveys state-of-the-art AI techniques addressing these challenges, and presents real-world case studies that demonstrate the practical benefits of AI-driven approaches. The chapter concludes with insights into future directions and best practices for integrating AI into SDN resource management.

Resource Allocation Challenges

Resource allocation in multi-tenant SDNs involves distributing limited network and compute resources—such as bandwidth, CPU cycles, memory, and storage—among tenants with diverse and often conflicting needs. The following subsections detail the primary challenges that complicate this task.

Dynamic and Heterogeneous Tenant Demands

Tenant workloads in multi-tenant SDNs are highly dynamic and heterogeneous. For example, one tenant may run latency-sensitive real-time applications such as video conferencing, while another may primarily process batch data analytics requiring high throughput but tolerating higher delay. Workload patterns can vary hourly, daily, or seasonally, influenced by user behavior, business cycles, and external events.

Static allocation schemes cannot adapt to these fluctuations, leading to resource underutilization during low demand and congestion or SLA violations during peak demand. AI-driven dynamic allocation is essential to continuously match resource supply with tenant demand.

Tenant Isolation and Security

Isolation is a fundamental requirement in multi-tenant environments to prevent tenants from interfering with each other's traffic or data. Resource allocation mechanisms must enforce strict isolation policies, ensuring that one tenant's resource consumption does not degrade others' performance or compromise security.

Moreover, isolation mechanisms must defend against resource contention attacks, where a malicious tenant attempts to exhaust shared resources to disrupt others. AI can help detect and mitigate such behavior by monitoring usage patterns and identifying anomalies.

Scalability and Real-Time Decision-Making

As the number of tenants and network devices grows, the scale and complexity of resource allocation decisions increase exponentially. Controllers must process large volumes of telemetry data—such as flow statistics, queue lengths, and CPU loads—and make allocation decisions within tight latency constraints to avoid network performance degradation.

Efficient AI algorithms that operate in real time and scale horizontally are necessary to meet these operational requirements. Techniques such as distributed learning and hierarchical control architectures are promising approaches.

Service Level Agreements (SLAs) and Quality of Service (QoS)

Multi-tenant SDN providers must comply with diverse SLAs specifying minimum bandwidth, maximum latency, jitter, and availability guarantees. These SLAs often conflict; for example, maximizing throughput may increase latency for certain tenants.

Resource allocation must balance these competing objectives, enforce priority policies, and provide fairness among tenants. Violations of SLAs can lead to financial penalties, reputational damage, and customer churn.

Resource Fragmentation and Utilization Efficiency

Resource fragmentation occurs when available resources are scattered in small, unusable blocks across the infrastructure. For instance, leftover bandwidth slices too small to satisfy any tenant's demand reduce overall utilization efficiency.

Optimizing resource packing and allocation to minimize fragmentation while respecting tenant constraints is a complex combinatorial problem. AI techniques such as heuristic search, genetic algorithms, and reinforcement learning can explore large solution spaces efficiently.

Multi-domain and Multi-layer Resource Coordination

In many SDN deployments, resources span multiple domains (e.g., data center, metro, and wide area networks) and multiple layers (e.g., physical, virtual, and application layers). Coordinating resource allocation across these heterogeneous domains and layers adds further complexity.

AI agents capable of multi-domain awareness and cross-layer optimization can provide end-to-end resource management that meets tenant requirements holistically.

AI Solutions for Multi-tenant Environments

Artificial intelligence offers a rich toolkit to tackle the aforementioned challenges. This section explores key AI techniques and their applications in multi-tenant SDN resource allocation.

Reinforcement Learning for Dynamic Resource Allocation

Reinforcement learning (RL) is well-suited for dynamic resource allocation, where decisions must adapt to evolving network states.

1. **Markov Decision Process (MDP) Formulation:** The resource allocation problem can be modeled as an MDP where states represent current network load, tenant demands, and resource availability; actions correspond to allocation decisions; and rewards reflect objectives like throughput, fairness, and SLA compliance.

2. **Deep Reinforcement Learning (DRL):** DRL combines RL with deep neural networks to handle high-dimensional state spaces. Algorithms such as Deep Q-Networks (DQN), Deep Deterministic Policy Gradient (DDPG), and Proximal Policy Optimization (PPO) have been applied successfully.

3. **Benefits:** RL agents learn optimal policies through interaction, adapting to changing conditions without explicit programming. They can balance multiple objectives and handle uncertainty.

4. **Challenges:** RL requires careful reward design, exploration–exploitation trade-offs, and training stability. Simulators or testbeds are often needed for training before deployment.

Predictive Analytics and Demand Forecasting

Accurate demand forecasting enables proactive resource provisioning, reducing SLA violations and resource wastage.

- **Time Series Models:** Traditional models like ARIMA and Holt–Winters capture seasonal and trend components in tenant usage data.

- **Machine Learning Models:** Support vector regression (SVR), random forests, and gradient boosting machines (GBM) can learn complex nonlinear patterns.

- **Deep Learning Models:** Recurrent neural networks (RNNs), especially Long Short-Term Memory (LSTM) networks, excel at capturing temporal dependencies and long-range correlations.

- **Application:** Forecasts inform reservation of bandwidth, CPU, and memory resources ahead of demand spikes, enabling smooth tenant experiences.

Multi-objective and Constraint-Aware Optimization

Resource allocation must optimize multiple conflicting objectives under constraints.

- **Evolutionary Algorithms:** Genetic algorithms and particle swarm optimization explore large solution spaces, balancing throughput, latency, fairness, and energy consumption.

- **Constraint Programming:** Ensures allocations respect tenant isolation, SLA thresholds, and resource limits.

- **Hybrid Approaches:** Combine heuristics with AI models to achieve near-optimal solutions with practical computation times.

AI-Based Traffic Engineering and Routing Optimization

Optimizing routing paths reduces congestion and improves QoS.

- **Graph Neural Networks (GNNs):** Model network topology and traffic flows, enabling AI to predict congestion points and recommend routing changes.

- **Deep Learning for Path Selection:** Neural networks learn to select optimal paths based on historical traffic and network states.

- **Reinforcement Learning for Routing:** RL agents adapt routing policies to dynamic traffic conditions.

Autonomous and Closed-Loop Resource Management

Integrated AI frameworks combine monitoring, prediction, optimization, and control in closed feedback loops.

- **Telemetry Collection:** AI agents continuously gather data from network devices and tenants.

- **Decision-Making:** AI models analyze data, forecast demand, and generate allocation decisions.

- **Actuation:** Controllers implement decisions by configuring network devices.

- **Feedback:** Performance metrics feed back into AI models for continuous learning and improvement.

This autonomy reduces human intervention, accelerates response times, and enhances reliability.

Table 11-1. *AI Techniques Applied in Multi-tenant SDN Resource Allocation*

AI Technique	Description	Application Example
Reinforcement Learning	Learns allocation policies through trial and error	Dynamic bandwidth allocation adapting to demand
Predictive Analytics	Forecasts future resource usage	Proactive CPU and memory provisioning
Multi-objective Optimization	Balances throughput, latency, and fairness	SLA-aware resource scheduling
AI-Based Traffic Engineering	Optimizes routing for load balancing	Congestion avoidance using network topology models
Autonomous Management	Integrates AI functions for end-to-end control	Self-optimizing multi-tenant SDN infrastructure

Case Studies in Resource Allocation

Real-world deployments illustrate how AI-driven resource allocation enhances multi-tenant SDN performance.

Case Study 1: Reinforcement Learning for Bandwidth Allocation in Cloud Data Centers

A leading cloud provider deployed a deep reinforcement learning system to dynamically allocate bandwidth among tenants sharing a data center network. The RL agent continuously monitored network congestion and tenant demand metrics, learning to adjust bandwidth shares in real time.

- **Implementation:** The system used a Deep Q-Network trained on simulated network traffic patterns, with reward functions balancing utilization and SLA compliance.

- **Outcomes:**

  - Achieved a 15% increase in average bandwidth utilization

  - Reduced SLA violations by 30%

  - Enabled automatic adaptation to workload spikes without manual tuning

- **Impact:** The approach improved tenant satisfaction and reduced operational costs by optimizing resource usage.

Case Study 2: Predictive Resource Provisioning in 5G Network Slicing

A telecom operator supporting 5G network slicing used machine learning models to forecast slice resource demands based on historical traffic and user mobility data.

- **Approach:** LSTM neural networks predicted bandwidth and compute needs for each slice 30 minutes ahead.

- **Results:**

  - Reduced resource wastage by 20%

  - Enhanced latency guarantees for ultra-reliable low-latency communication (URLLC) slices

  - Improved SLA compliance and tenant experience

- **Significance:** Proactive provisioning enabled efficient resource sharing while meeting diverse slice requirements.

Case Study 3: AI-Driven Traffic Engineering in Enterprise SDN

An enterprise network deployed an AI-based traffic engineering solution using graph neural networks (GNNs) to model network topology and traffic flows.

- **Method:** The GNN predicted congestion hotspots, and a reinforcement learning agent adjusted routing paths dynamically.

- **Benefits:**

  - Increased network throughput by 18%

  - Reduced average packet delay by 22%

  - Maintained strict tenant isolation without manual configuration

- **Conclusion:** AI-enabled routing optimization significantly enhanced network performance and tenant satisfaction.

AI-Driven Resource Allocation Architecture (Enhanced)

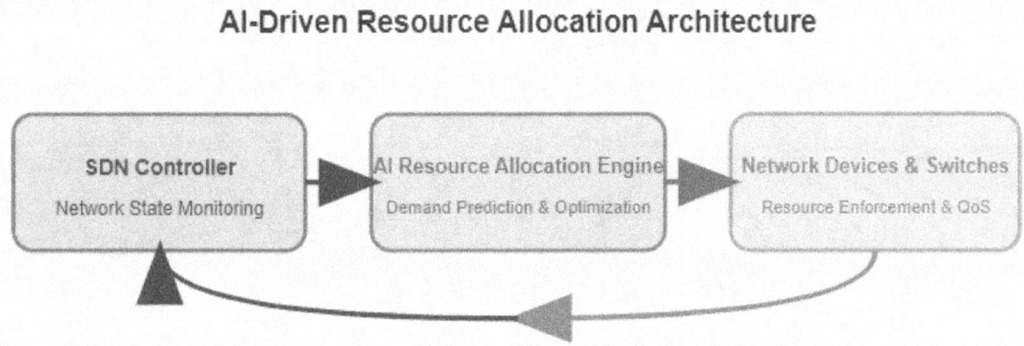

Figure 11-1. *AI-Driven Resource Allocation Architecture*

This diagram illustrates the closed-loop interaction between the SDN controller, AI resource allocation engine, and network devices in a multi-tenant SDN. The AI engine leverages monitoring data to predict demand and optimize resource allocation, which is enforced by network devices to meet tenant SLAs.

Conclusion

Resource allocation in multi-tenant SDN environments is a critical and challenging task due to the dynamic, heterogeneous, and multidimensional nature of tenant demands and network resources. Achieving efficient, fair, and secure resource distribution requires adaptive, scalable, and intelligent mechanisms.

Artificial intelligence provides a versatile and powerful set of tools—ranging from reinforcement learning and predictive analytics to multi-objective optimization and AI-based traffic engineering—that enable dynamic, proactive, and autonomous resource management. These AI-driven approaches improve resource utilization, ensure SLA compliance, and enhance tenant satisfaction.

The case studies presented demonstrate that AI-based resource allocation is not merely theoretical but is already delivering measurable benefits in cloud data centers, telecom networks, and enterprise SDNs. As SDN technologies evolve and multi-tenant environments become more complex, integrating AI into resource allocation frameworks will be essential for building resilient, scalable, and intelligent networks.

Network operators and researchers should continue to explore hybrid AI models, federated learning for privacy-preserving collaboration, and explainable AI techniques to increase transparency and trust in automated resource management systems.

CHAPTER 12

AI Agents for Edge-Centric SDN Management

The convergence of Software-Defined Networking (SDN) and edge computing represents a significant paradigm shift in modern network architecture. As computational resources move closer to data sources and end users, traditional centralized network management approaches face new challenges in scalability, latency, and autonomy. Edge-centric SDN environments distribute intelligence across the network edge, enabling localized decision-making while maintaining centralized oversight.

This distributed architecture creates unique management complexities that conventional rule-based systems struggle to address. Artificial intelligence (AI) agents offer promising solutions by bringing adaptive, autonomous decision-making capabilities to edge nodes. These AI agents can learn from local conditions, make real-time decisions, and coordinate with other agents to optimize network performance and resource utilization.

This chapter explores the intersection of edge computing, SDN, and artificial intelligence. We examine the fundamental concepts of edge computing in SDN contexts, analyze the critical roles AI plays in edge management, and present compelling use cases that demonstrate the transformative potential of AI agents in edge-centric networks. Through this exploration, we aim to provide network architects and operators with insights into leveraging AI for more efficient, responsive, and intelligent edge-centric SDN deployments.

© Het Mehta 2026
H. Mehta, *AI Agents for Secure and Software-Defined Networking*,
https://doi.org/10.1007/979-8-8688-2358-9_12

Edge Computing Overview

Edge computing fundamentally transforms network architecture by distributing computational resources closer to data sources and users, reducing latency and bandwidth consumption while improving service responsiveness.

Edge Computing Fundamentals

Edge computing refers to the deployment of computing and storage resources at or near the network edge, in contrast to traditional centralized cloud or data center models. This paradigm addresses several limitations of centralized architectures:

- **Latency Reduction**: By processing data closer to its source, edge computing minimizes round-trip delays, enabling time-sensitive applications like autonomous vehicles, industrial automation, and augmented reality.

- **Bandwidth Optimization**: Local data processing reduces the volume of information transmitted to central locations, alleviating backbone network congestion and lowering bandwidth costs.

- **Reliability Enhancement**: Edge nodes can continue functioning during central network outages, improving service availability and resilience.

- **Privacy Preservation**: Sensitive data can be processed locally, reducing exposure to potential security breaches during transmission.

Edge computing exists on a continuum from device-level edge (embedded systems, IoT devices) through local edge (gateways, on-premises servers) to regional edge (micro data centers, telecom facilities), each layer offering different computational capabilities and latency characteristics.

SDN in Edge Environments

Software-Defined Networking provides the programmability and flexibility needed to manage complex edge environments effectively:

- **Hierarchical Control Planes**: Edge-centric SDN typically implements hierarchical control structures with global controllers for policy management and local controllers for tactical decisions.

- **Dynamic Resource Allocation**: SDN enables dynamic allocation of network resources based on application requirements, user demands, and edge node capabilities.

- **Service Function Chaining**: SDN facilitates the creation and management of service chains across distributed edge resources, ensuring optimal data flow paths.

- **Network Slicing**: Edge-centric SDN supports network slicing to isolate resources for different applications or tenants, crucial for multi-tenant edge deployments.

Challenges in Edge-Centric SDN Management

Despite its advantages, edge-centric SDN introduces significant management challenges:

- **Heterogeneity**: Edge environments encompass diverse hardware platforms, connectivity options, and computational capabilities, complicating uniform management.

- **Scale**: The sheer number of edge nodes—potentially thousands or millions—overwhelms traditional management approaches.

- **Dynamism**: Edge environments experience frequent changes in connectivity, workload, and resource availability, requiring adaptive management solutions.

- **Limited Resources**: Edge nodes often have constrained computational, storage, and energy resources compared to cloud data centers.

- **Partial Connectivity**: Intermittent or unreliable connectivity between edge nodes and central controllers necessitates autonomous operation capabilities.

These challenges highlight the need for intelligent, autonomous management solutions that can operate effectively within the constraints of edge environments while delivering optimal performance.

AI's Role in Edge Management

Artificial intelligence transforms edge-centric SDN management by enabling autonomous decision-making, predictive optimization, and adaptive resource allocation across distributed environments.

Distributed AI Architectures

AI deployment in edge environments requires architectures that balance local autonomy with global coordination:

- **Federated Learning**: Enables edge nodes to collaboratively train AI models while keeping data local, preserving privacy and reducing bandwidth requirements. Models are trained locally, and only parameter updates are shared with central coordinators.

- **Hierarchical AI**: Implements multilevel AI decision-making where simple, time-sensitive decisions occur at the edge while complex, resource-intensive analytics happen at higher levels.

- **Multi-agent Systems**: Deploys autonomous AI agents across edge nodes that communicate and coordinate to achieve global optimization while making independent local decisions.

These distributed architectures allow AI to function effectively despite the constraints and heterogeneity of edge environments.

Autonomous Network Management

AI agents enable autonomous management of edge-centric SDN environments:

- **Self-Configuration**: AI agents automatically configure network parameters based on local conditions, application requirements, and available resources.

- **Self-Healing**: Agents detect failures or performance degradation and implement remediation actions without human intervention.

- **Self-Optimization**: Continuous monitoring and adjustment of network parameters optimize performance metrics like latency, throughput, and energy efficiency.

- **Anomaly Detection**: AI models identify unusual patterns that may indicate security threats, hardware failures, or application issues.

This autonomy is crucial for edge environments where manual management is impractical due to scale and geographic distribution.

Predictive Resource Management

AI enables proactive resource management through predictive capabilities:

- **Workload Forecasting**: Machine learning models predict future computational and network demands based on historical patterns, enabling preemptive resource allocation.

- **User Mobility Prediction**: AI anticipates user movement patterns to optimize service placement and handover operations in mobile edge computing scenarios.

- **Predictive Maintenance**: AI identifies potential hardware or software failures before they occur, scheduling maintenance to minimize service disruption.

These predictive capabilities transform reactive management approaches into proactive optimization strategies.

Edge-Centric SDN Architecture with AI Agents

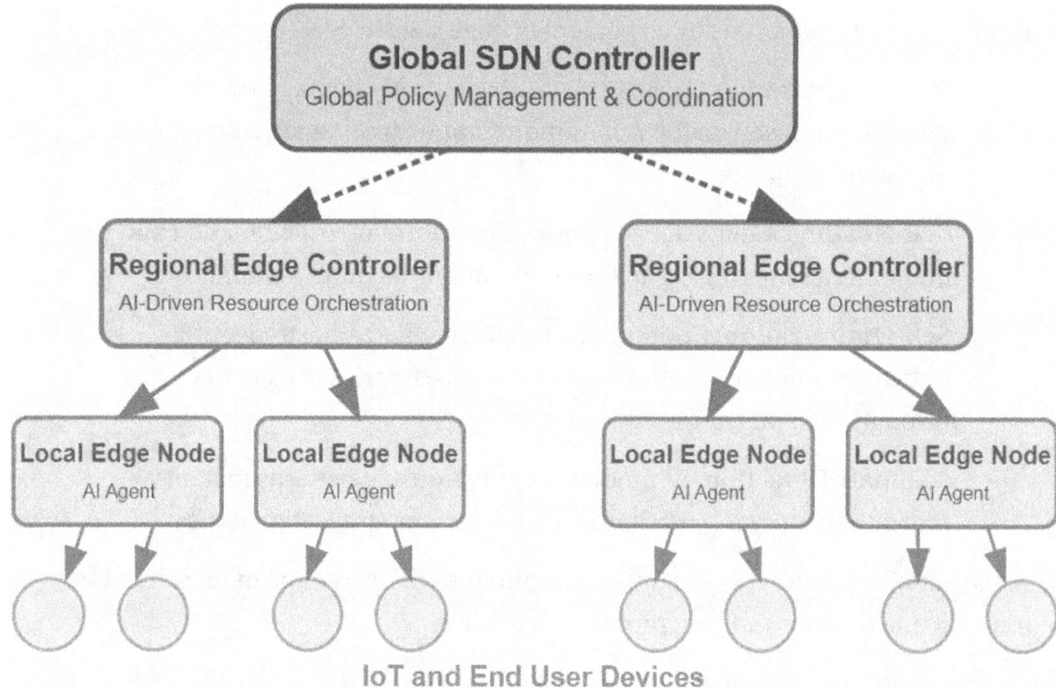

Figure 12-1. *Edge-Centric SDN Architecture with AI Agents*

This diagram illustrates the hierarchical nature of edge-centric SDN with AI agents deployed at different levels. The global SDN controller provides policy management and coordination, regional edge controllers handle resource orchestration, and local edge nodes with AI agents directly manage end devices. This multitier architecture enables both global optimization and local autonomy.

Context-Aware Decision-Making

AI agents leverage contextual awareness to make intelligent decisions:

- **Environmental Awareness**: Agents consider factors like time of day, location, weather conditions, and local events that may impact network demand.

- **Application Awareness**: AI models understand application requirements and characteristics to optimize resource allocation for different application types.

- **User Experience Awareness**: Decisions incorporate user experience metrics, prioritizing resources for quality-sensitive applications.

This contextual intelligence enables more nuanced and effective management decisions than traditional rule-based approaches.

Use Cases in Edge-Centric Networks

The integration of AI agents into edge-centric SDN enables transformative applications across multiple domains.

Smart Cities and Urban Infrastructure

AI-powered edge networks support intelligent urban services:

- **Traffic Management**: Edge nodes process real-time traffic camera feeds, with AI agents dynamically adjusting traffic signals to optimize flow based on current conditions.

- **Public Safety**: Distributed AI agents analyze surveillance footage locally, identifying potential security incidents while preserving privacy by only transmitting alerts rather than raw video.

- **Environmental Monitoring**: Edge-deployed AI analyzes sensor data to detect air quality issues, flooding risks, or other environmental concerns, triggering automated responses.

These applications benefit from edge processing to reduce latency and bandwidth consumption while enabling autonomous operation during connectivity disruptions.

Industrial IoT and Smart Manufacturing

Edge AI transforms industrial environments:

- **Predictive Maintenance**: AI agents at the factory edge analyze equipment sensor data to predict failures before they occur, scheduling maintenance during planned downtime.

- **Quality Control**: Computer vision models deployed on edge servers inspect products in real-time, identifying defects without sending sensitive production data to external clouds.

- **Process Optimization**: AI agents continuously monitor production parameters, automatically adjusting settings to optimize yield, quality, and energy efficiency.

The combination of edge computing and AI enables real-time decision-making critical for industrial applications with stringent latency and reliability requirements.

V2X (Vehicle-to-Everything)

Connected vehicles rely on edge AI for safe, efficient operation:

- **Cooperative Perception**: Vehicles share sensor data through edge nodes, with AI agents fusing information to create comprehensive environmental awareness beyond individual vehicle capabilities.

- **Dynamic Route Optimization**: Edge-deployed AI continuously recalculates optimal routes based on real-time traffic, weather, and road conditions.

- **Vehicle-to-Infrastructure Coordination**: AI agents manage interactions between vehicles and infrastructure elements like traffic signals, optimizing traffic flow and safety.

Edge computing reduces the latency critical for vehicle safety, while AI provides the intelligence needed for complex decision-making.

Table 12-1. *AI Agent Types in Edge-Centric SDN*

Agent Type	Primary Function	Deployment Location	Key Capabilities
Resource Allocation Agent	Optimizes resource distribution	Edge nodes	Workload prediction, priority-based allocation, resource efficiency
Traffic Engineering Agent	Manages network flows	Edge routers	Path optimization, congestion avoidance, QoS enforcement
Security Agent	Detects and mitigates threats	All edge tiers	Anomaly detection, threat identification, automated response
Service Placement Agent	Determines optimal service location	Regional controllers	Latency analysis, resource availability assessment, mobility prediction
Energy Management Agent	Optimizes power consumption	Edge nodes	Workload consolidation, sleep scheduling, thermal management

Augmented Reality and Immersive Experiences

Edge AI enables responsive, context-aware AR applications:

- **Low-Latency Rendering**: Edge nodes perform compute-intensive rendering tasks, streaming only the results to lightweight AR devices.

- **Context-Aware Content**: AI agents analyze the user's environment in real time, delivering relevant augmented content based on location, objects in view, and user preferences.

- **Collaborative AR**: Edge servers coordinate shared AR experiences among multiple users in proximity, maintaining consistent virtual elements across different devices.

These applications require both the low latency of edge computing and the intelligent processing capabilities of AI to deliver compelling user experiences.

Conclusion

AI agents are transforming edge-centric SDN management by enabling autonomous, intelligent decision-making at all levels of the network hierarchy. The distributed nature of edge computing demands new approaches to network management that can operate effectively despite constraints in connectivity, resources, and scale.

Through distributed AI architectures, autonomous network management capabilities, and predictive resource optimization, AI agents provide the intelligence needed to handle the complexity of edge environments. Real-world applications across smart cities, industrial IoT, autonomous transportation, and augmented reality demonstrate the practical value of this integration.

As edge computing continues to evolve, AI agents will play an increasingly central role in ensuring efficient, reliable, and responsive network operations. The future of edge-centric SDN lies in the development of more sophisticated AI techniques that can deliver even greater autonomy, adaptation, and intelligence across distributed network environments.

Organizations implementing edge-centric SDN should prioritize AI integration in their architectural planning, focusing on hierarchical control structures that balance local autonomy with global coordination. By leveraging the power of AI agents, network operators can unlock the full potential of edge computing while managing its inherent complexity.

CHAPTER 13

AI-Augmented Network Function Chaining in SDN

In the evolving landscape of modern networks, flexibility and agility have become paramount. Network Function Virtualization (NFV) and Software-Defined Networking (SDN) have revolutionized how network services are deployed and managed, enabling dynamic, programmable, and cost-effective infrastructures. One of the critical enablers within this framework is Network Function Chaining (NFC), which allows network operators to define and orchestrate ordered sequences of Virtualized Network Functions (VNFs) to deliver complex services tailored to specific application needs.

However, as networks grow in scale and complexity—especially with the advent of 5G, IoT, and edge computing—the manual design and management of function chains become increasingly challenging. Artificial intelligence (AI) offers transformative potential to augment network function chaining by automating chain composition, optimizing resource allocation, and adapting dynamically to changing network conditions.

This chapter delves into the fundamentals of NFV and network function chaining, explores the AI techniques that enhance chaining efficiency and intelligence, and examines future trends shaping this critical area of SDN-based network management.

Network Function Virtualization Basics

Understanding AI-augmented function chaining requires a solid grasp of Network Function Virtualization (NFV) and its role within SDN.

H. Mehta, *AI Agents for Secure and Software-Defined Networking*,
https://doi.org/10.1007/979-8-8688-2358-9_13

What Is Network Function Virtualization?

Network Function Virtualization is a paradigm shift from traditional hardware-based network appliances to software-based virtualized functions running on commodity servers. Instead of deploying dedicated physical devices for firewalls, load balancers, intrusion detection systems, and other network functions, NFV enables these functions to be instantiated as software instances—called Virtualized Network Functions (VNFs)—that can be dynamically deployed, scaled, and managed.

Key Benefits of NFV:

- **Flexibility:** VNFs can be deployed anywhere in the network and scaled up or down based on demand.

- **Cost Efficiency:** Reduces reliance on expensive proprietary hardware.

- **Faster Service Deployment:** New services can be rolled out rapidly by chaining VNFs in software.

- **Resource Optimization:** Shared infrastructure allows better utilization of compute and network resources.

Network Function Chaining Explained

Network Function Chaining (NFC), also known as Service Function Chaining (SFC), refers to the process of linking multiple VNFs in a specific, ordered sequence to form a complete network service. For example, a service chain might consist of a firewall, followed by a load balancer, then a deep packet inspection function.

The chaining process involves

- **Defining the Chain:** Specifying which VNFs to include and in what order

- **Traffic Steering:** Ensuring packets traverse the VNFs according to the chain

- **Resource Allocation:** Assigning compute, storage, and network resources to each VNF

- **Monitoring and Management:** Tracking performance, detecting faults, and dynamically adjusting the chain as needed

Challenges in Network Function Chaining

While NFC offers tremendous benefits, it also introduces several challenges:

- **Complexity of Chain Design:** Manually designing optimal chains for diverse services and tenants is complex and error-prone.

- **Resource Constraints:** VNFs compete for limited physical resources, requiring intelligent allocation to avoid bottlenecks.

- **Dynamic Network Conditions:** Traffic patterns and service demands fluctuate, necessitating adaptive chaining.

- **Latency and Performance Guarantees:** Chains must meet strict latency and throughput requirements, especially for real-time applications.

- **Security and Isolation:** Ensuring that chained VNFs do not introduce vulnerabilities or interfere with other tenants' traffic.

These challenges motivate the integration of AI techniques to automate and optimize network function chaining.

AI Techniques for Function Chaining

Artificial intelligence brings powerful capabilities to address the complexities of network function chaining. By leveraging machine learning, optimization algorithms, and intelligent agents, AI can automate chain design, enhance resource utilization, and adapt to real-time network conditions.

AI for Automated Chain Composition

Traditionally, network operators manually define function chains based on predefined templates or heuristics. AI enables automated chain composition by learning from historical data and network policies.

- **Reinforcement Learning (RL):** RL agents interact with the network environment to learn optimal chaining policies that maximize performance metrics such as throughput and minimize latency. By receiving feedback from the environment, RL can adapt chaining strategies dynamically.

- **Supervised Learning:** Classification and regression models can predict the best sequence of VNFs for specific traffic types or applications based on labeled datasets.

- **Graph-Based Models:** Graph Neural Networks (GNNs) are particularly essential here, as network topology and VNF dependencies form inherently graph-structured data. GNNs can model complex inter-VNF communication patterns and resource dependencies that linear models cannot capture.

AI-Driven Resource Allocation and Scaling

Efficient resource management is critical for maintaining service quality in function chains. AI techniques optimize resource allocation across VNFs to balance load and reduce latency.

- **Predictive Analytics:** Machine learning models forecast traffic loads and resource demands, enabling proactive scaling of VNFs before congestion occurs.

- **Optimization Algorithms:** Genetic algorithms, particle swarm optimization, and other metaheuristic methods search for near-optimal resource allocation configurations that satisfy multiple constraints, including CPU, memory, and bandwidth.

- **Deep Reinforcement Learning:** DRL agents continuously monitor network state and adjust resource allocation in real time, learning policies that optimize chain performance under varying conditions.

Dynamic Traffic Steering and Load Balancing

AI enhances traffic steering mechanisms to ensure packets traverse VNFs efficiently within the chain.

- **Traffic Classification:** AI models classify traffic flows by type and priority, enabling differentiated treatment in chaining.

- **Adaptive Routing:** Reinforcement learning techniques optimize routing paths to avoid congestion and reduce latency.

- **Load Balancing:** AI agents distribute traffic evenly across multiple VNF instances, preventing overload and improving fault tolerance.

Fault Detection and Self-Healing

AI agents embedded within the chaining framework can detect faults and initiate corrective actions autonomously.

- **Anomaly Detection:** Unsupervised learning algorithms identify deviations from normal VNF behavior, signaling potential failures or attacks.

- **Predictive Maintenance:** AI predicts VNF degradation or impending failures, allowing preemptive migration or restart.

- **Automated Recovery:** Self-healing mechanisms triggered by AI can reroute traffic or spin up new VNF instances to maintain chain continuity.

AI-Augmented Network Function Chaining Architecture

This figure illustrates how user traffic is steered through a sequence of VNFs such as firewall, load balancer, and deep packet inspection (DPI). An AI agent oversees chain composition, resource allocation, and traffic optimization, enabling dynamic and intelligent chaining.

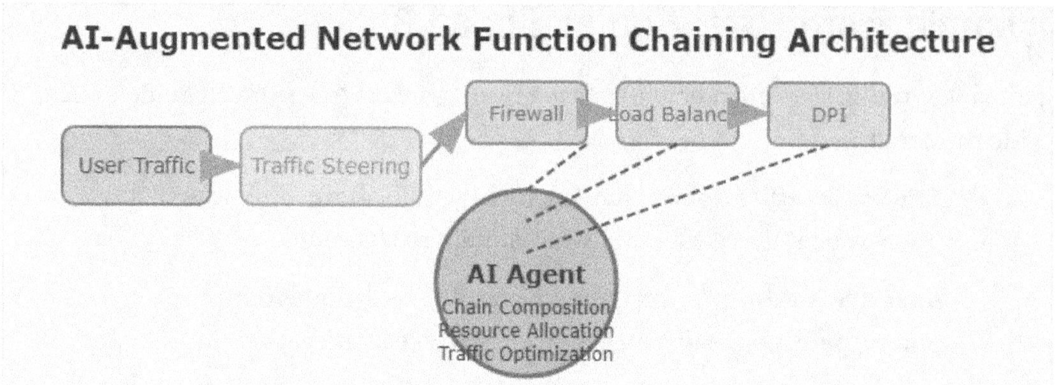

Figure 13-1. *AI-Augmented Network Function Chaining Architecture*

Case Study: AI-Augmented Function Chaining in 5G Networks

In 5G networks, network slicing relies heavily on dynamic function chaining to provide tailored services for diverse applications such as enhanced mobile broadband, ultra-reliable low-latency communication, and massive IoT.

An AI-augmented chaining system uses reinforcement learning to

- Automatically compose chains optimized for slice-specific SLAs

- Predict traffic surges and scale VNFs proactively

- Adapt traffic steering to minimize latency and avoid congestion

This approach has demonstrated significant improvements in resource utilization and SLA compliance in experimental 5G testbeds.

Future Trends in Function Chaining

The integration of AI with network function chaining is an evolving field, with several promising trends shaping its future.

Explainable AI for Transparent Chain Management

As AI-driven chaining decisions impact critical network services, transparency becomes essential. Explainable AI (XAI) techniques will enable network operators to understand and trust AI recommendations by providing interpretable models and decision rationales.

This transparency will facilitate troubleshooting, compliance with regulations, and operator confidence in automated systems.

Federated and Collaborative AI Models

Given the distributed nature of SDN and NFV environments, federated learning approaches will enable multiple edge and cloud nodes to collaboratively train AI models without sharing sensitive data.

Such collaborative AI will enhance chain optimization across administrative domains and improve privacy preservation.

Integration with Digital Twins

Digital twins—virtual replicas of physical network environments—will be increasingly used to simulate and test function chaining strategies before deployment.

AI agents will leverage digital twins to evaluate chaining policies, predict performance impacts, and optimize configurations in a risk-free environment.

Multi-domain and Cross-Layer Chaining

Future networks will extend chaining beyond a single administrative domain or network layer, requiring AI to orchestrate VNFs across multiple domains and layers (e.g., physical, virtual, application).

AI-driven cross-layer orchestration will enable end-to-end service guarantees, seamless mobility, and holistic resource optimization.

Energy-Efficient Function Chaining

With growing environmental concerns, AI will play a crucial role in designing energy-aware function chains that minimize power consumption without compromising performance.

Techniques such as workload consolidation, dynamic VNF placement, and energy-aware routing will be guided by AI models balancing efficiency and service quality.

Table 13-1. *AI Techniques and Their Roles in Network Function Chaining*

AI Technique	Primary Role	Benefits	Example Use Case
Reinforcement Learning	Dynamic chain composition and routing	Adapts to changing network conditions	SLA-based chaining in 5G network slices
Supervised Learning	VNF sequence prediction	Quick classification of traffic types	Automated chain design for enterprise apps
Graph Neural Networks	Modeling network dependencies	Captures complex VNF relationships	Optimizing multi-domain chaining
Predictive Analytics	Forecasting resource demand	Enables proactive scaling	Traffic surge prediction for VNFs
Anomaly Detection	Fault and security event detection	Early fault identification and mitigation	Self-healing VNF chains
Optimization Algorithms	Resource allocation	Efficient utilization of compute/network resources	Load balancing across VNFs

Conclusion

Network Function Chaining is a cornerstone technology enabling flexible, programmable, and efficient network services in SDN and NFV environments. However, the increasing complexity and dynamic nature of modern networks demand intelligent automation to design, manage, and optimize function chains effectively.

Artificial intelligence offers a comprehensive toolkit—from reinforcement learning and predictive analytics to graph neural networks and anomaly detection—that significantly enhances network function chaining. AI-driven automation reduces operational complexity, improves resource utilization, ensures SLA compliance, and enables self-healing capabilities.

As AI techniques continue to mature, future developments will focus on explainability, collaborative learning, digital twin integration, multi-domain orchestration, and energy efficiency. These advances will further empower network operators to deliver agile, reliable, and sustainable services.

In summary, AI-augmented network function chaining represents a vital evolution in SDN management, poised to meet the demands of next-generation networks and services.

CHAPTER 14

AI Agents for Zero-Touch Provisioning in SDN

The rapid expansion of modern networks, fueled by cloud computing, the Internet of Things (IoT), 5G, and edge computing, has created unprecedented complexity in network management. Traditional manual configuration and provisioning approaches are no longer scalable or efficient enough to meet the demands of these dynamic environments. Network operators face the challenge of deploying, configuring, and managing thousands or even millions of network devices and services with minimal errors and delays.

Zero-touch provisioning (ZTP) emerges as a critical solution to this problem. ZTP automates the entire lifecycle of network device onboarding, configuration, software updates, and activation, eliminating the need for manual intervention. This automation not only accelerates network deployment but also reduces operational costs and human errors, which are common causes of network outages.

Software-Defined Networking (SDN), with its centralized control plane and programmable interfaces, provides a powerful platform to implement ZTP frameworks. The programmability and visibility offered by SDN controllers enable fine-grained control over device provisioning workflows, making it possible to automate complex network tasks.

However, the sheer scale and complexity of modern networks, combined with heterogeneous devices and dynamic traffic patterns, require intelligent automation that goes beyond static scripts and rule-based systems. Artificial intelligence (AI) agents bring cognitive capabilities such as learning, reasoning, and adaptation, empowering ZTP solutions to become truly autonomous, context-aware, and self-optimizing.

© Het Mehta 2026
H. Mehta, *AI Agents for Secure and Software-Defined Networking*,
https://doi.org/10.1007/979-8-8688-2358-9_14

This chapter explores the foundational concepts of zero-touch provisioning, examines how AI techniques enhance automation in SDN environments, and discusses the key challenges and future directions in deploying AI-driven ZTP solutions. Through this exploration, readers will gain a comprehensive understanding of how AI agents are transforming network provisioning and management.

Zero-Touch Provisioning Overview

To begin, let's first understand the fundamental concept behind zero-touch provisioning.

What Is Zero-Touch Provisioning?

Zero-touch provisioning (ZTP) is an automation paradigm designed to enable network devices to configure themselves automatically when connected to the network, with minimal or no human intervention. Upon powering on and connecting to the network, a device initiates a process to discover the network environment, authenticate itself securely, download necessary software images and configuration files, and activate its services.

The traditional approach to provisioning involves network engineers manually configuring each device with specific parameters, which is time-consuming, error-prone, and difficult to scale. ZTP addresses these limitations by automating key provisioning phases:

1. **Discovery and Authentication:** As soon as the device connects, it identifies itself to the network using methods such as DHCP options, secure enrollment protocols, or device certificates. Authentication ensures that only authorized devices gain access.

2. **Configuration Delivery:** The device downloads its configuration files, firmware, and software updates from centralized repositories or SDN controllers. Configurations can include IP addresses, routing policies, security parameters, and service-specific settings.

3. **Activation and Verification:** After applying configurations, the device activates its functions and performs self-tests. It reports its operational status back to the management system to confirm successful provisioning.

This automated workflow reduces provisioning time from hours or days to minutes, enabling rapid network scaling and dynamic service deployment.

Role of SDN in Zero-Touch Provisioning

Software-Defined Networking (SDN) plays a pivotal role in enabling and enhancing zero-touch provisioning. SDN's fundamental principle of separating the control plane from the data plane and centralizing network intelligence into a programmable controller provides several advantages for ZTP:

- **Centralized Orchestration:** The SDN controller maintains a global view of the network topology, device inventory, and policies. It orchestrates device onboarding and provisioning workflows centrally, ensuring consistency and coordination across the network.

- **Programmability and APIs:** SDN exposes northbound APIs that allow automation systems and AI agents to programmatically control network devices, push configurations, and monitor status in real time.

- **Real-Time Telemetry and Monitoring:** Continuous monitoring of device health, traffic patterns, and network events enables dynamic adjustments during provisioning, such as retrying failed configurations or scaling resources.

- **Multi-vendor Interoperability:** SDN abstracts underlying hardware differences, allowing ZTP workflows to operate uniformly across heterogeneous devices from different vendors.

- **Policy Enforcement:** Centralized policy engines within SDN controllers ensure that provisioning adheres to organizational security and compliance requirements.

Together, SDN and ZTP form a synergistic combination that transforms network deployment from a manual, error-prone process into an agile, automated, and reliable operation.

Benefits of Zero-Touch Provisioning

The adoption of zero-touch provisioning in SDN environments yields multiple benefits:

- **Operational Efficiency:** Automating repetitive provisioning tasks frees network engineers from manual configurations, allowing them to focus on strategic activities such as network design and optimization.

- **Accelerated Time to Market:** New devices and services can be brought online rapidly, supporting business agility and faster rollout of innovations.

- **Consistency and Accuracy:** Automated workflows ensure uniform configurations across devices, reducing inconsistencies that lead to network failures or security vulnerabilities.

- **Scalability:** ZTP supports massive device deployments typical in data centers, enterprise campuses, and IoT ecosystems, enabling networks to grow without proportional increases in operational overhead.

- **Reduced Human Errors:** Manual configurations are a major source of network outages. ZTP minimizes these errors by standardizing and automating provisioning steps.

- **Cost Savings:** Reduced labor costs, fewer outages, and faster deployments translate to significant operational savings.

In summary, ZTP is a foundational capability for modern networks that require agility, scale, and reliability.

AI Methods for Automation

While traditional zero-touch provisioning frameworks rely on static scripts, predefined templates, and rule-based workflows, these approaches lack the flexibility and intelligence needed for highly dynamic and complex networks. Artificial intelligence (AI) introduces cognitive capabilities that enable ZTP systems to learn from data, adapt to changing conditions, and make autonomous decisions.

This section explores the key AI techniques that enhance zero-touch provisioning in SDN.

AI-Driven Device Discovery and Authentication

Device discovery and authentication are critical first steps in the provisioning process. AI agents improve these stages by analyzing traffic patterns, device signatures, and behavioral profiles to accurately identify and authenticate devices joining the network.

- **Machine Learning for Device Fingerprinting:** Supervised learning models classify devices based on network traffic characteristics, protocol usage, and hardware fingerprints, enabling precise identification even in heterogeneous environments.

- **Anomaly Detection for Security:** Unsupervised learning algorithms detect unusual patterns during device onboarding that may indicate unauthorized access attempts, rogue devices, or compromised hardware.

- **Behavioral Profiling:** AI builds dynamic profiles of legitimate devices, learning their typical behavior over time. Deviations from these profiles trigger alerts or additional authentication steps.

By leveraging AI, networks can improve security and reduce false positives during device onboarding.

Intelligent Configuration Generation

Traditional ZTP systems often use static configuration templates that cannot adapt to varying network contexts or device roles. AI agents can dynamically generate optimized configurations tailored to the device's function, network conditions, and service requirements.

- **Natural Language Processing (NLP):** AI can interpret high-level network policies or operator intents expressed in natural language and translate them into device-specific configuration commands. This reduces the need for manual coding of configurations.

- **Reinforcement Learning (RL):** RL agents interact with the network environment to learn optimal configuration parameters that maximize performance metrics such as throughput, latency, and reliability. Over time, RL adapts configurations based on feedback.

- **Generative Models:** Techniques such as Generative Adversarial Networks (GANs) can create novel configuration sets that optimize network performance or security posture under varying conditions.

- **Context-Aware Configuration:** AI considers real-time telemetry data, such as traffic load or fault conditions, to adjust configurations dynamically during provisioning.

This intelligent configuration generation makes provisioning more flexible, efficient, and resilient.

Automated Software Image Management

Managing software versions, patches, and updates across large fleets of devices is a complex task. AI assists by predicting optimal timing for updates, selecting compatible software images, and minimizing service disruptions.

- **Predictive Maintenance:** AI models analyze device usage patterns, error logs, and performance metrics to forecast when software updates or patches should be applied before failures occur.

- **Compatibility Analysis:** Machine learning algorithms assess hardware capabilities, dependencies, and network policies to recommend compatible software images, avoiding version conflicts.

- **Update Scheduling:** AI optimizes the timing of software rollouts to minimize impact on network performance and service availability, considering factors like peak traffic periods.

Automating software image management reduces downtime and improves network stability.

Context-Aware Provisioning and Adaptation

AI agents leverage real-time network telemetry, environmental data, and historical trends to adapt provisioning workflows dynamically.

- **Traffic Prediction:** Time-series forecasting models predict future traffic loads, enabling AI to provision additional resources proactively or delay noncritical provisioning during congestion.

- **Policy Conflict Detection and Resolution:** AI detects conflicts between provisioning policies, such as overlapping IP address assignments or incompatible security rules, and suggests or applies resolutions automatically.

- **Multi-domain Coordination:** In multi-administrative domain networks, AI orchestrates provisioning workflows across different segments to ensure seamless end-to-end service activation.

- **Environmental Awareness:** AI considers external factors such as energy consumption targets or regulatory constraints when making provisioning decisions.

This context-aware adaptation enhances provisioning efficiency and compliance.

Closed-Loop Automation with AI

One of the most powerful applications of AI in zero-touch provisioning is enabling closed-loop automation, where provisioning actions are continuously monitored and adjusted based on feedback from the network environment.

- **Continuous Learning:** AI models update themselves with new telemetry and operational data, improving provisioning accuracy and adapting to evolving network conditions.

- **Self-Healing:** Upon detecting provisioning failures or suboptimal configurations, AI triggers corrective actions such as rolling back changes, retrying configurations, or reallocating resources.

- **Performance Optimization:** AI fine-tunes provisioning parameters to meet service level agreements (SLAs), optimize resource usage, and maintain network stability.

- **Root Cause Analysis:** AI analyzes provisioning failures to identify underlying causes, enabling proactive prevention in future deployments.

Closed-loop AI automation transforms provisioning from a linear process into a resilient, adaptive system.

AI-Driven Zero-Touch Provisioning Framework

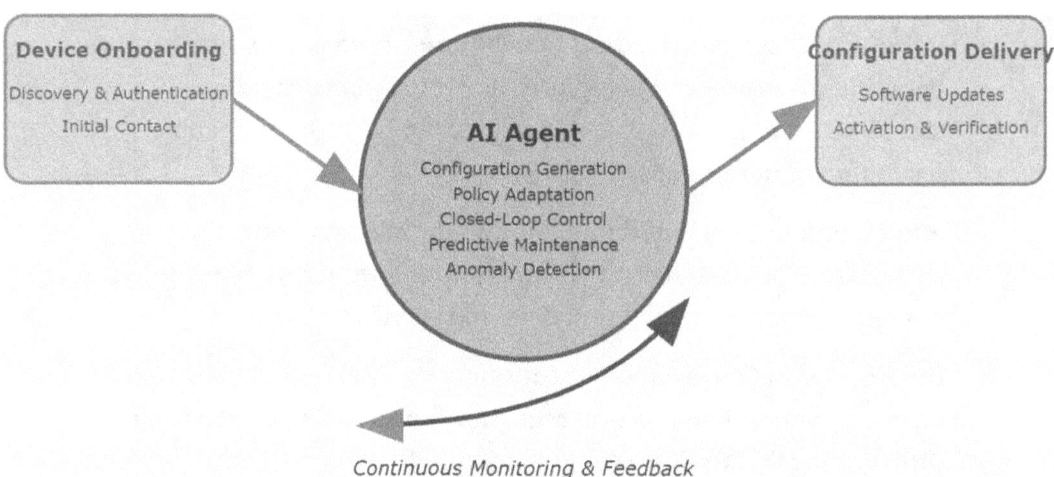

Figure 14-1. *AI-Driven Zero-Touch Provisioning Framework*

This figure depicts the core components of AI-enhanced zero-touch provisioning. Devices onboard the network, AI agents generate and adapt configurations, and provisioning is delivered seamlessly with continuous feedback and closed-loop control.

Challenges in Zero-Touch Provisioning

Despite its transformative potential, AI-driven zero-touch provisioning faces several technical and operational challenges that must be addressed to ensure robust, secure, and scalable deployments.

Data Quality and Availability

AI models depend heavily on the availability of high-quality, representative data for training, validation, and real-time decision-making. Many networks face challenges related to data quality:

- **Incomplete and Noisy Data:** Telemetry streams may have missing values, inconsistent formats, or measurement errors, which degrade AI model accuracy.

- **Data Silos:** Data is often fragmented across different management systems, vendors, and administrative domains, limiting holistic insights.

- **Labeling Challenges:** Supervised learning requires labeled datasets, but generating accurate labels for network events or configurations is labor-intensive and costly.

- **Real-Time Processing Requirements:** Provisioning decisions often require near real-time data ingestion and processing, demanding efficient data pipelines and low-latency analytics.

Addressing these data challenges involves investing in robust data collection frameworks, preprocessing techniques, and federated learning approaches that enable collaborative model training without sharing sensitive raw data.

Security and Privacy Concerns

Automating provisioning introduces new attack surfaces and risks:

- **Unauthorized Access and Rogue Devices:** Automated onboarding processes must include strong authentication and authorization mechanisms to prevent malicious devices from infiltrating the network.

- **Model Poisoning Attacks:** Adversaries may attempt to manipulate training data or telemetry to corrupt AI models, leading to incorrect provisioning decisions.

- **Sensitive Data Exposure:** AI models processing network and user data must comply with privacy regulations such as GDPR, requiring encryption, anonymization, and access controls.

- **Supply Chain Risks:** Firmware and software updates delivered automatically must be verified to prevent injection of malicious code.

Robust security frameworks, including secure enclaves for AI execution, adversarial training, and continuous monitoring, are essential to mitigate these risks.

Interoperability and Standardization

Modern networks are heterogeneous, composed of devices from multiple vendors with varying capabilities, protocols, and interfaces.

- **Vendor-Specific APIs:** Lack of standardized APIs complicates AI-driven provisioning workflows, requiring custom adapters and integrations.

- **Legacy Devices:** Many existing devices lack support for automated provisioning protocols such as NETCONF or RESTCONF, limiting ZTP applicability.

- **Protocol Compatibility:** Ensuring AI agents can operate seamlessly across diverse protocols and device types is critical for end-to-end automation.

Industry efforts towards standardization, such as the IETF Service Function Chaining (SFC) standards and OpenConfig models, are vital to improving interoperability and simplifying AI integration.

Scalability and Performance

Deploying AI agents for zero-touch provisioning at scale involves balancing computational overhead with provisioning speed and responsiveness.

- **Resource Constraints:** AI models must be optimized to run efficiently on SDN controllers or edge devices with limited CPU, memory, and power.

- **Latency Sensitivity:** Delays in provisioning can impact service availability, especially in time-critical applications like industrial control or autonomous vehicles.

- **Distributed Coordination:** In multi-domain or multicloud environments, coordinating AI agents and synchronizing provisioning actions require robust communication and consensus mechanisms.

Techniques such as model compression, edge computing, and hierarchical AI architectures help address scalability and performance challenges.

Trust and Human Oversight

Despite advances in AI, network operators may be hesitant to fully trust autonomous provisioning systems without transparency and control mechanisms.

- **Explainability:** AI decisions must be interpretable to enable operators to understand, validate, and trust provisioning actions.

- **Fallback and Override:** Manual intervention options should be available to override AI decisions in case of failures or unexpected situations.

- **Policy Compliance:** AI agents must ensure provisioning actions comply with organizational policies, regulatory requirements, and security standards.

Developing explainable AI (XAI) techniques, interactive dashboards, and hybrid human–AI workflows will be critical for building operator confidence.

Table 14-1. *Challenges in AI-Driven Zero-Touch Provisioning and Potential Solutions*

Challenge	Description	Potential AI or System Solutions
Data Quality	Incomplete, noisy, or siloed data	Data preprocessing, federated learning, anomaly detection
Security and Privacy	Unauthorized access, model poisoning, data leaks	Secure authentication, adversarial training, encryption
Interoperability	Vendor heterogeneity, legacy system limitations	Standardization efforts, protocol abstraction layers
Scalability	Computational overhead, latency constraints	Lightweight AI models, edge computing, distributed AI
Trust and Oversight	Operator reluctance, lack of explainability	Explainable AI (XAI), transparent dashboards, manual override

Conclusion

Zero-touch provisioning is revolutionizing network management by automating device onboarding, configuration, and activation, thereby enabling scalable, efficient, and error-free network operations. When integrated with the programmability and centralized control of SDN, ZTP becomes a powerful enabler of network agility and innovation.

Artificial intelligence agents significantly enhance zero-touch provisioning by introducing adaptability, context-awareness, predictive capabilities, and closed-loop control. AI-driven ZTP systems can autonomously discover devices, generate optimized configurations, manage software updates, and continuously adapt to changing network conditions, all while maintaining security and compliance.

However, realizing the full potential of AI-powered zero-touch provisioning requires overcoming challenges related to data quality, security, interoperability, scalability, and operator trust. Addressing these challenges through robust AI models, industry-standard protocols, secure frameworks, and transparent interfaces will pave the way for widespread adoption.

Looking forward, AI agents will be indispensable in managing increasingly complex and dynamic networks, supporting rapid service innovation while ensuring reliability, security, and operational excellence.

CHAPTER 15

Leveraging Graph Neural Networks for SDN Management

The management of modern networks, especially Software-Defined Networks (SDN), is becoming increasingly complex due to the scale, heterogeneity, and dynamic nature of network topologies and traffic patterns. Traditional machine learning and optimization techniques often struggle to capture the intricate relationships and dependencies inherent in network structures. Graph Neural Networks (GNNs), a cutting-edge deep learning paradigm designed to operate directly on graph-structured data, offer unprecedented capabilities to model and analyze network topologies, device interactions, and traffic flows.

This chapter introduces the fundamental concepts of Graph Neural Networks, explores their diverse applications in SDN management, and discusses emerging trends that are shaping the future of graph-based approaches in networking. By harnessing the power of GNNs, network operators and researchers can unlock new levels of automation, intelligence, and efficiency in SDN environments.

Introduction to Graph Neural Networks

Let's start by exploring the basic idea and significance of graph neural networks.

© Het Mehta 2026
H. Mehta, *AI Agents for Secure and Software-Defined Networking*,
https://doi.org/10.1007/979-8-8688-2358-9_15

What Are Graph Neural Networks?

Graph Neural Networks (GNNs) are a class of neural networks designed to process data represented as graphs, where entities are modeled as nodes and their relationships as edges. Unlike traditional neural networks that operate on fixed-size vectors or grids (e.g., images), GNNs can learn from the complex, non-Euclidean structures typical of graphs, making them ideal for representing social networks, molecular structures, knowledge graphs, and, importantly, communication networks.

At their core, GNNs iteratively aggregate and transform information from a node's neighbors to learn rich node embeddings that capture both local and global graph structure. This message-passing mechanism allows GNNs to model dependencies and interactions effectively.

Key Components and Architecture

A typical GNN architecture consists of several layers, each performing a *message passing* step:

- **Node Embeddings:** Each node starts with an initial feature vector representing its attributes (e.g., device type, capacity).

- **Message Passing:** At each layer, nodes exchange information with their neighbors via edges. Messages are aggregated using functions such as sum, mean, or max.

- **Update Function:** The aggregated message is combined with the node's current embedding using neural network functions (e.g., MLPs) to produce updated embeddings.

- **Readout Layer:** After multiple layers, a readout function aggregates node embeddings to produce graph-level outputs or predictions.

Common variants of GNNs include Graph Convolutional Networks (GCNs), Graph Attention Networks (GATs), and GraphSAGE, each differing in how messages are aggregated and weighted.

Why GNNs Are Suitable for Network Management

Network infrastructures naturally form graph structures: switches, routers, and hosts are nodes; links and communication paths are edges. This graph representation captures crucial topological and relational information that traditional vector-based models cannot easily exploit.

GNNs excel in

- **Modeling Network Topology:** Learning representations that encode connectivity, link quality, and device roles

- **Capturing Dependencies:** Representing how failures or performance issues propagate through the network

- **Generalization:** Applying learned models to unseen network configurations due to their inductive learning capabilities

- **Multi-scale Analysis:** Combining local neighborhood information with global network context

These properties make GNNs uniquely suited to address complex SDN management tasks such as fault diagnosis, traffic prediction, routing optimization, and security analysis.

Applications in SDN Management

Transitioning from theory to practice, this section explores how Graph Neural Networks are applied to enhance various aspects of SDN management, enabling smarter, more adaptive networks.

Traffic Prediction and Flow Classification

Accurate prediction of network traffic and classification of flows are essential for efficient resource allocation and congestion avoidance.

- **Graph-Based Traffic Prediction:** GNNs model the network as a graph where nodes represent switches or routers and edges represent links. By learning from historical traffic data and topological features, GNNs predict traffic volume and flow patterns on links or nodes.

- **Flow Classification:** GNNs incorporate both flow-level features and relational information such as flow dependencies and shared paths to classify traffic types (e.g., video, VoIP, bulk data), enabling differentiated service treatment.

Compared to traditional time-series or feature-based models, GNNs leverage spatial correlations in the network, improving prediction accuracy and robustness.

Fault Detection and Localization

Network reliability depends on the timely detection and localization of faults such as link failures, misconfigurations, or hardware malfunctions.

- **Anomaly Detection:** GNNs analyze network telemetry represented as graphs to detect anomalies in node behavior or link performance, considering the impact on neighboring components.

- **Fault Localization:** By learning propagation patterns of failures through the network graph, GNNs can pinpoint the root cause of observed symptoms, reducing troubleshooting time.

This graph-aware approach outperforms traditional methods that analyze nodes or links in isolation.

Routing Optimization and Path Selection

Optimizing routing decisions is critical to maximize throughput, minimize latency, and maintain network resilience.

- **GNN-Based Routing Policies:** GNNs learn embeddings of network states that inform routing algorithms about congestion, link reliability, and path diversity.

- **Dynamic Path Selection:** By predicting the impact of routing changes on network performance, GNNs assist SDN controllers in selecting optimal paths adaptively.

- **Multi-objective Optimization:** GNNs can encode multiple criteria (e.g., delay, bandwidth, energy consumption) to support balanced routing decisions.

This leads to more efficient and adaptive routing compared to static or heuristic-based approaches.

Security and Intrusion Detection

Network security benefits significantly from graph-based analysis, as attacks often manifest as complex patterns involving multiple devices and paths.

- **Intrusion Detection Systems (IDS):** GNNs analyze communication graphs to identify suspicious patterns such as lateral movement, distributed denial of service (DDoS) attacks, or botnet activity.

- **Malware Propagation Modeling:** By learning how malware spreads through network graphs, GNNs predict vulnerable nodes and recommend mitigation strategies.

- **Access Control and Policy Verification:** GNNs verify consistency and compliance of network policies by modeling access relationships.

Graph-based security analytics provide deeper insights than flat feature-based methods, improving detection rates and reducing false positives.

Network Resource Allocation and Virtual Network Embedding

Efficient allocation of physical resources to Virtualized Network Function is crucial in SDN-enabled NFV environments.

- **Virtual Network Embedding (VNE):** GNNs model both physical and virtual networks as graphs and learn embeddings to map virtual nodes and links onto physical substrates optimally.

- **Resource Prediction:** GNNs forecast resource demands by analyzing network usage patterns and topology, aiding proactive scaling.

- **Load Balancing:** By understanding relational dependencies, GNNs guide traffic distribution to avoid hotspots.

These applications improve resource utilization and service quality.

Graph Neural Networks Applications in SDN Management

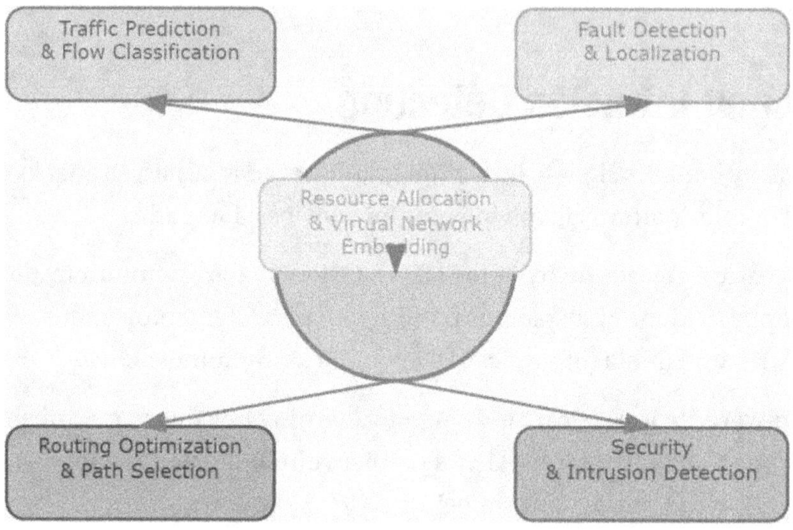

Figure 15-1. *Applications of Graph Neural Networks in SDN Management*

This diagram illustrates the key areas where GNNs contribute to SDN management, including traffic prediction, fault detection, routing optimization, security, and resource allocation.

Case Study: GNNs for Traffic Engineering in SDN

In a recent research initiative, GNNs were applied to optimize traffic engineering in a large-scale SDN. The network was modeled as a graph with switches as nodes and links as edges, enriched with real-time traffic and link quality metrics.

The GNN model learned to predict link utilization and identify bottlenecks, enabling the SDN controller to proactively reroute flows. This approach reduced average latency by 15% and improved throughput by 20% compared to traditional heuristic methods.

This case study demonstrates the practical benefits of integrating GNNs into SDN control frameworks.

Future Trends in Graph-Based Approaches

As research and industry adoption of Graph Neural Networks in SDN management accelerate, several emerging trends and challenges are shaping the future landscape.

Explainable and Interpretable GNNs

One major challenge with deep learning models, including GNNs, is their black-box nature. For network operators, understanding why a GNN makes a particular decision is crucial for trust and troubleshooting.

Future research is focusing on developing explainable GNN models that provide insights into which nodes, edges, or features influenced predictions. Techniques such as attention visualization, gradient-based attribution, and rule extraction are promising approaches.

Explainability will increase operator confidence and facilitate regulatory compliance.

Integration with Digital Twins and Simulation

Digital twins—virtual replicas of physical networks—enable safe testing and validation of network management strategies.

Combining GNNs with digital twins allows training and evaluating graph-based models in simulated environments before deployment. This integration supports what-if analyses, scenario planning, and risk assessment.

It also accelerates innovation by reducing reliance on live network experimentation.

Multi-layer and Multi-domain Graph Modeling

Future networks span multiple layers (physical, virtual, application) and administrative domains.

Extending GNNs to model multi-layer graphs that capture interdependencies across layers will enable holistic network management. Similarly, multi-domain graph models will facilitate end-to-end orchestration and fault management across organizational boundaries.

This requires scalable GNN architectures and cross-domain data sharing frameworks.

Real-Time and Streaming GNNs

Network management demands real-time responsiveness to dynamic events.

Developing GNN models capable of processing streaming data and updating embeddings incrementally is an active research area. Streaming GNNs will enable timely detection of anomalies, rapid adaptation to topology changes, and continuous optimization.

Efficient algorithms and hardware acceleration will be key enablers.

Federated and Privacy-Preserving GNNs

Privacy concerns and regulatory requirements often restrict sharing raw network data across domains.

Federated learning approaches, where GNN models are trained collaboratively without exchanging raw data, are gaining traction. Privacy-preserving techniques such as differential privacy and secure multiparty computation will enable broader adoption of GNNs in sensitive environments. These methods balance model accuracy with data confidentiality.

Conclusion

Graph Neural Networks represent a powerful and versatile tool for advancing SDN management. Their ability to naturally model complex network topologies and relational data enables improved traffic prediction, fault diagnosis, routing optimization, security analytics, and resource allocation.

This chapter provided a comprehensive introduction to GNNs, highlighted their practical applications in SDN, and explored future trends shaping graph-based network intelligence. As networks continue to grow in scale and complexity, leveraging GNNs will be essential to build adaptive, resilient, and efficient SDN infrastructures.

Investing in explainable models, integrating with digital twins, supporting multi-layer and multi-domain graphs, and addressing real-time and privacy challenges will unlock the full potential of graph-based approaches in next-generation network management.

CHAPTER 16

AI-Enhanced Security for Programmable Data Planes

The evolution of programmable data planes has transformed modern networking by enabling unprecedented flexibility and high-speed packet processing directly within network devices. Technologies such as P4-programmable switches allow network operators to define custom packet-processing pipelines, implement fine-grained telemetry, and accelerate network functions at hardware speeds. While this programmability brings powerful capabilities, it also introduces new security challenges that were previously nonexistent or limited in traditional fixed-function data planes.

The programmable data plane's dynamic and complex nature increases the attack surface, exposes new vulnerabilities, and complicates real-time threat detection. Conventional security mechanisms, designed for static environments, struggle to keep pace with these challenges. Artificial intelligence (AI), with its ability to analyze vast amounts of data, detect subtle anomalies, and automate adaptive responses, offers promising solutions to secure programmable data planes effectively.

This chapter explores the security challenges unique to programmable data planes, reviews AI techniques that enhance security, discusses future research directions, and highlights AI's emerging role in certificate and key management within these environments. By understanding and leveraging AI-enhanced security, network operators can build resilient, intelligent, and trustworthy programmable network infrastructures.

© Het Mehta 2026
H. Mehta, *AI Agents for Secure and Software-Defined Networking*,
https://doi.org/10.1007/979-8-8688-2358-9_16

Security Challenges in Data Planes

Programmable data planes introduce a paradigm shift in network architecture, but this shift comes with significant security implications. Understanding these challenges is critical to designing effective AI-driven defenses.

Expanded Attack Surface Due to Programmability

Traditional data planes are implemented using fixed-function ASICs with well-understood and limited behavior. Programmable data planes, however, execute custom code defined by network operators, often written in languages like P4. This flexibility creates a broader attack surface:

- **Code Vulnerabilities:** Malicious actors or accidental bugs in P4 programs can introduce vulnerabilities such as buffer overflows, logic errors, or backdoors.

- **Unauthorized Code Injection:** Attackers may exploit software update mechanisms or supply chain weaknesses to inject unauthorized or malicious code into data plane devices.

- **Control Interface Exploits:** The interfaces used to program and manage data planes can be targeted to manipulate device behavior, bypassing traditional control-plane security.

This expanded surface requires continuous validation, verification, and monitoring of programmable code and update processes.

Insider Threats and Misconfigurations

The flexibility of programmable data planes increases the risk of insider threats and configuration errors:

- **Insider Attacks:** Trusted personnel with programming access might insert malicious logic, intentionally or unintentionally, to disrupt network operations or leak data.

- **Misconfigurations:** Complex programmable pipelines increase the likelihood of configuration errors, which can lead to traffic leaks, policy violations, or denial of service.

- **Complex Control-Data Plane Interactions:** The interplay between control plane instructions and data plane programs complicates security audits and anomaly detection.

AI-driven monitoring can help detect suspicious changes and misconfigurations early.

Real-Time Threat Detection Challenges

Programmable data planes operate at line rate, processing millions of packets per second. This imposes stringent requirements for real-time security:

- **Limited Visibility:** The high-speed nature limits the ability to inspect every packet deeply without impacting performance.

- **Telemetry Overload:** Programmable devices generate vast amounts of telemetry data, which must be efficiently analyzed to detect threats.

- **False Positives and Latency:** Balancing sensitivity and specificity in threat detection is critical to avoid unnecessary disruptions or missed attacks.

AI techniques can analyze high-dimensional telemetry data in near real-time, identifying subtle anomalies indicative of attacks.

Secure Key and Certificate Management

Secure communications within programmable data planes depend on robust cryptographic key and certificate management:

- **Key Generation and Storage:** Ensuring keys are generated securely and stored in tamper-resistant hardware.

- **Rotation and Revocation:** Regularly rotating keys and revoking compromised certificates without disrupting network operations.

- **Automation:** Managing certificates and keys at scale in dynamic environments requires automation to avoid human errors.

AI can assist by predicting optimal rotation schedules, detecting anomalies in usage, and automating lifecycle management.

AI Techniques for Enhancing Security

Artificial intelligence provides powerful tools to address the multifaceted security challenges of programmable data planes.

Anomaly Detection Using Machine Learning

Machine learning (ML) models analyze telemetry data such as packet headers, flow statistics, and device logs to detect deviations from normal behavior:

- **Unsupervised Learning:** Techniques like clustering and autoencoders learn normal patterns without requiring labeled attack data, enabling detection of novel threats.

- **Supervised Learning:** Classification models trained on labeled datasets identify known attack signatures with high accuracy.

- **Semi-supervised Learning:** Combines both approaches, leveraging limited labeled data and abundant unlabeled data to improve detection.

By modeling spatial and temporal correlations in network traffic, ML algorithms can identify anomalies such as distributed denial of service (DDoS), spoofing, or data exfiltration attempts.

Deep Learning for Intrusion Detection

Deep learning models, including Recurrent Neural Networks (RNNs) and Convolutional Neural Networks (CNNs), capture complex temporal and spatial patterns in network data:

- **RNNs:** Analyze sequences of packets or flows to detect sophisticated intrusion attempts that unfold over time.

- **CNNs:** Extract features from traffic matrices or graph representations of network interactions, identifying subtle attack patterns.

These models reduce false positives and improve detection of advanced persistent threats (APTs).

Reinforcement Learning for Automated Response

Reinforcement learning (RL) agents interact with the network environment to learn optimal security policies:

- **Automated Mitigation:** RL agents learn when and how to apply mitigation actions such as traffic filtering, rerouting, or quarantining suspicious flows.

- **Adaptation:** Policies evolve based on feedback, enabling proactive defense against emerging threats.

- **Resource Optimization:** RL balances security actions with network performance constraints.

This approach enables dynamic, context-aware security enforcement without human intervention.

Explainable AI for Security Transparency

Security teams require transparent AI models to trust and effectively use automated systems:

- **Feature Attribution:** Identifies which inputs influenced AI decisions, helping analysts understand alerts.

- **Visualization:** Graphical explanations of detected threats and affected network components improve situational awareness.

- **Rule Extraction:** Converts complex AI models into human-readable rules to support compliance and audits.

Explainable AI (XAI) bridges the gap between AI automation and human oversight.

Graph Neural Networks for Network Security

Graph Neural Networks (GNNs) model network topology and communication patterns as graphs, enabling advanced security analytics:

- **Attack Propagation Modeling:** GNNs learn how attacks spread across devices and links.

- **Anomaly Detection:** Detect irregular communication patterns that indicate lateral movement or botnet activity.

- **Policy Verification:** Analyze access control graphs to identify policy violations.

GNNs complement other AI models by incorporating relational information critical in programmable networks.

Future Directions in Security

The intersection of AI and programmable data planes is a rapidly evolving area with several promising future directions.

SmartNICs

Future programmable data planes will embed AI accelerators directly within network devices:

- **Line-Rate Detection:** Enables real-time threat detection and mitigation without offloading to external controllers.

- **Reduced Latency:** Immediate responses to threats reduce damage and improve resilience.

- **Energy Efficiency:** Specialized hardware optimizes AI inference power consumption.

This integration will transform programmable data planes into intelligent security enforcement points.

Collaborative AI Security Frameworks

Distributed AI agents deployed across network elements will collaborate to share threat intelligence:

- **Federated Learning:** Enables training global AI models without exchanging sensitive raw data.

- **Coordinated Defense:** Agents communicate to correlate anomalies and orchestrate network-wide responses.

- **Threat Intelligence Sharing:** Real-time exchange of indicators of compromise (IoCs) enhances detection accuracy.

Collaborative frameworks enhance security posture across multi-domain and multi-tenant environments.

Adaptive and Context-Aware Security Policies

AI will enable continuous adaptation of security policies based on

- **Network Context:** Traffic patterns, topology changes, and device states

- **User Behavior:** Dynamic risk profiles based on user activities

- **Threat Landscape:** Emerging vulnerabilities and attack trends

Adaptive policies improve security effectiveness while minimizing operational impact.

Privacy-Preserving AI Techniques

Privacy concerns limit data sharing for AI security analysis:

- **Federated Learning:** Collaborative model training without raw data exchange

- **Homomorphic Encryption:** Enables computations on encrypted data

- **Differential Privacy:** Protects individual data points in shared datasets

These techniques will enable broader adoption of AI-driven security in sensitive environments.

AI for Certificate and Key Management

Cryptographic keys and certificates underpin secure communications in programmable data planes. AI can significantly enhance their management.

Challenges in Key and Certificate Lifecycle Management

Managing cryptographic materials in dynamic programmable environments involves

- **Dynamic Provisioning:** Frequent addition/removal of network devices requires automated key distribution.

- **Key Rotation:** Regular rotation minimizes the risk of key compromise but complicates management.

- **Certificate Revocation:** Timely revocation is critical to prevent misuse of compromised certificates.

- **Heterogeneous Devices:** Different hardware capabilities and security requirements complicate uniform management.

Manual management is error-prone and unscalable.

AI-Driven Automation for Lifecycle Management

AI models can automate and optimize key and certificate management by:

- **Predictive Analytics:** Forecasting optimal rotation intervals based on usage patterns and threat intelligence

- **Anomaly Detection:** Identifying unusual certificate usage or key access indicating potential compromise

- **Workflow Automation:** Orchestrating certificate issuance, renewal, and revocation with minimal human intervention

- **Risk-Based Authentication:** Evaluating trustworthiness of devices and certificates using behavioral analytics

Automation improves security posture and operational efficiency.

Enhancing Trust and Compliance

AI can support compliance with security standards by continuously monitoring cryptographic asset health and generating audit reports. Risk scores derived from AI models help prioritize remediation efforts.

AI-Enhanced Security for Programmable Data Planes Conceptual Overview

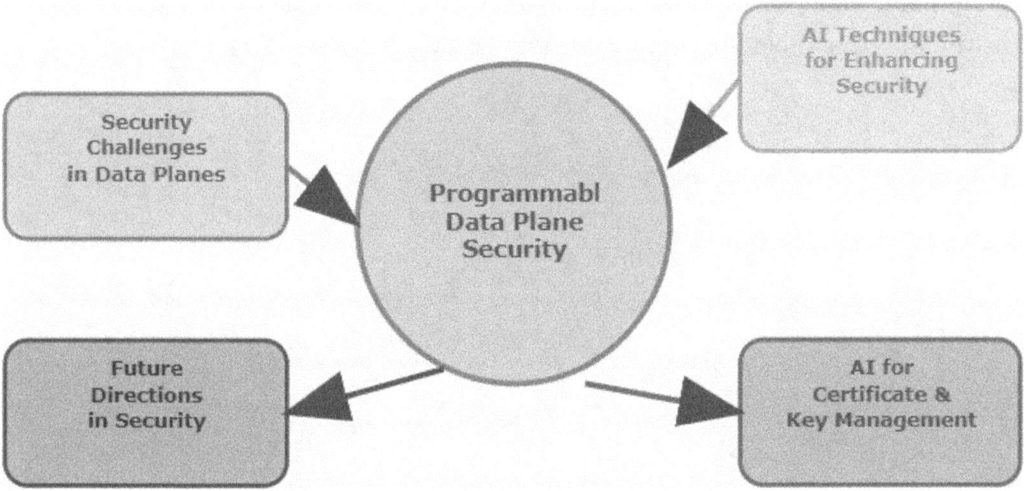

Figure 16-1. *Conceptual Overview of AI-Enhanced Security for Programmable Data Planes*

This figure illustrates the core concept: security challenges feed into programmable data plane security, which is enhanced by AI techniques. The outcomes include future security directions and AI-driven certificate and key management.

Conclusion

Programmable data planes are a cornerstone of modern network innovation, delivering flexibility and performance but also introducing novel security challenges. The expanded attack surface, insider risks, real-time detection difficulties, and cryptographic management complexities require advanced, intelligent solutions.

Artificial intelligence offers a comprehensive toolkit to enhance programmable data plane security. From machine learning-based anomaly detection and deep learning intrusion detection to reinforcement learning for automated response and explainable AI for transparency, these technologies empower networks to defend themselves adaptively and proactively.

Looking ahead, embedding AI capabilities directly into programmable hardware, fostering collaborative AI security frameworks, developing adaptive policies, and ensuring privacy-preserving AI applications will be critical. Furthermore, AI-driven automation in certificate and key management will streamline security operations and strengthen trust.

By embracing AI-enhanced security, network operators can safeguard programmable data planes, ensuring resilient, efficient, and trustworthy network infrastructures that meet the demands of tomorrow's digital world.

CHAPTER 17

Personalized Network Management with AI Agents

The rapid proliferation of connected devices, diverse applications, and ever-changing user demands has transformed the landscape of network management. Traditional network management approaches, which typically rely on static configurations and uniform policies applied broadly across users and services, are increasingly inadequate in addressing the unique, dynamic requirements of modern network environments. The concept of personalized network management has emerged as a critical paradigm shift, focusing on tailoring network behavior, policies, and resource allocation to the specific needs, preferences, and context of individual users or user groups.

Personalized network management aims to optimize network performance, security, and user satisfaction by dynamically adapting to the heterogeneity inherent in today's networks. This approach not only enhances service quality but also improves operational efficiency by aligning network resources more closely with actual user demands.

Artificial intelligence (AI) plays a pivotal role in enabling personalized network management at scale. Through continuous learning from data, predictive modeling, and autonomous decision-making, AI agents can anticipate user needs, adapt network configurations in real time, and provide intuitive interfaces for user interaction. This chapter explores the foundations of personalization in networking, delves into AI methodologies that facilitate customization, and examines the profound impact of personalization on user experience and satisfaction.

© Het Mehta 2026
H. Mehta, *AI Agents for Secure and Software-Defined Networking,*
https://doi.org/10.1007/979-8-8688-2358-9_17

Personalization in Networking

To appreciate personalization in networking, it's important to first recognize why the demand for it is increasing.

The Growing Need for Personalization

Modern networks serve an increasingly diverse set of users, devices, and applications, each with distinct characteristics and performance expectations. For instance:

- **Mobile Users:** Devices such as smartphones and tablets frequently move across different network cells and environments, causing fluctuations in connectivity quality and bandwidth availability. Personalized management can optimize handoffs, bandwidth allocation, and latency reduction based on user mobility patterns.

- **Internet of Things (IoT) Devices:** IoT devices range from low-power sensors to smart appliances, each with unique communication patterns and security needs. For example, a medical sensor requires high reliability and strict security, while a smart thermostat may tolerate lower QoS.

- **Enterprise Applications:** Business-critical applications like video conferencing, financial transactions, and cloud services demand guaranteed QoS levels, low latency, and strong security policies.

- **Multimedia Streaming:** Video and audio streaming services require smooth playback with minimal buffering and adaptive bitrate adjustments tailored to user device capabilities and network conditions.

In such a heterogeneous environment, applying uniform policies leads to suboptimal network utilization and poor user experiences. Personalization addresses this by

- Allocating resources dynamically to meet individual user or application requirements

- Prioritizing traffic flows based on user preferences, subscription levels, or criticality

- Adapting security policies to the risk profile and trustworthiness of users and devices

- Adjusting network configurations in response to real-time context such as location, time of day, or device type

By doing so, personalization enhances network efficiency, user satisfaction, and overall service quality.

Dimensions and Scope of Personalization

Personalization in networking spans multiple dimensions, each addressing different aspects of network behavior:

Traffic Management

Traffic management involves controlling how data packets are handled within the network. Personalization allows

- **Differentiated Traffic Prioritization:** For example, prioritizing voice-over-IP (VoIP) packets for users who frequently make calls while deprioritizing background software updates

- **Dynamic Traffic Shaping:** Adjusting bandwidth limits and shaping policies based on user activity patterns or subscription tiers

- **Context-Aware Routing:** Selecting optimal paths based on user location, device type, and application sensitivity

Quality of Service (QoS) Customization

QoS parameters such as latency, jitter, packet loss, and bandwidth are critical for application performance. Personalization enables

- **Adaptive QoS Profiles:** Tailoring QoS settings for individual users or applications, e.g., providing low-latency paths for gamers or high-throughput for video streamers

- **Service Level Agreement (SLA) Enforcement:** Ensuring personalized SLAs are met according to user contracts or policies

Security Policy Adaptation

Security is a paramount concern in network management. Personalized security involves

- **User-Specific Access Controls:** Granting or restricting access based on user roles, device trust levels, or behavioral profiles

- **Anomaly Detection Sensitivity:** Adjusting intrusion detection system thresholds based on user risk profiles to reduce false positives

- **Contextual Authentication:** Applying multifactor authentication selectively depending on user location or device

Resource Allocation

Efficient resource utilization is essential for cost-effective network operations. Personalization supports

- **Dynamic Bandwidth Allocation:** Allocating bandwidth based on predicted user demand and priority

- **Edge Computing Resource Management:** Assigning compute and storage resources near the user dynamically to reduce latency

- **Load Balancing:** Personalized load distribution across servers or network paths to optimize performance

Service Function Chaining (SFC)

SFC defines the sequence of network functions (e.g., firewall, DPI, NAT) that packets traverse. Personalization allows

- **Custom SFC Paths:** Different user groups or applications can have tailored service chains optimized for security and performance.

- **Dynamic SFC Reconfiguration:** Adjusting chains in real time based on user context or threat intelligence.

Challenges in Implementing Personalization

While the benefits of personalization are clear, implementing it at scale presents significant challenges:

- **Data Collection and Privacy:** Personalization relies on collecting detailed user and network data, raising privacy concerns and regulatory compliance issues (e.g., GDPR, CCPA). Balancing data utility with privacy protection is critical.

- **Scalability and Complexity:** Managing personalized policies for millions of users requires scalable architectures and efficient algorithms to avoid overwhelming network controllers.

- **Real-Time Adaptation:** Networks must respond swiftly to changes in user context and network conditions without causing service disruptions.

- **Interoperability:** Personalized management systems must integrate seamlessly with diverse network devices, protocols, and legacy systems.

- **User Control and Transparency:** Users should have control over their personalization settings and understand how their data is used.

Addressing these challenges necessitates intelligent automation and robust AI-driven frameworks.

AI Approaches for Customization

Artificial intelligence provides the foundation for scalable, intelligent, and adaptive personalized network management. AI techniques automate complex decision-making, learn from data, and enable proactive network optimization.

User Profiling and Behavior Modeling

Building accurate user profiles is fundamental to personalization. AI leverages diverse data sources including traffic logs, device metadata, application usage, and user feedback.

- **Clustering Techniques:** Unsupervised learning algorithms such as k-means, DBSCAN, and hierarchical clustering group users with similar network behaviors or preferences. For example, clustering might reveal groups of users who predominantly use streaming services vs. those focused on real-time communications.

- **Classification Models:** Supervised learning models categorize users into predefined classes based on labeled historical data. These classes could represent business vs. consumer users, high vs. low bandwidth consumers, or trusted vs. risky devices.

- **Sequence and Time-Series Modeling:** Recurrent Neural Networks (RNNs), Long Short-Term Memory (LSTM) networks, and Hidden Markov Models (HMMs) analyze temporal sequences of user actions or network events to predict future behavior. Such models enable proactive resource allocation and preemptive issue resolution.

- **Contextual Awareness:** AI models incorporate contextual data such as location, device type, time of day, and social context to refine user profiles.

These profiles inform policy decisions, enabling the network to anticipate and meet user needs dynamically.

Reinforcement Learning for Dynamic Adaptation

Reinforcement learning (RL) is a powerful paradigm for learning optimal policies through interaction with the environment. In personalized network management:

- **Agent–Environment Interaction:** The RL agent observes network states (e.g., congestion levels, user QoS feedback) and takes actions (e.g., adjusting routing, changing bandwidth allocations) to maximize cumulative rewards related to user satisfaction and network efficiency.

- **Exploration vs. Exploitation:** RL balances exploring new configurations to discover better policies and exploiting known good policies to maintain performance.

- **Multi-objective Optimization:** RL can optimize multiple conflicting objectives such as minimizing latency while maximizing fairness and security.

RL agents have been applied to personalized traffic routing, adaptive resource scheduling in cloud-edge environments, and dynamic security policy enforcement.

RL enables networks to adapt continuously to evolving user demands and network conditions without explicit reprogramming.

Natural Language Processing for User Interaction

Natural Language Processing (NLP) enhances personalization by enabling intuitive communication between users and network management systems:

- **Conversational Interfaces:** Voice assistants and chatbots allow users to specify preferences, report issues, or request changes without technical expertise.

- **Intent Recognition:** NLP models parse user commands to extract intents and parameters, translating natural language into actionable network policies.

- **Sentiment and Feedback Analysis:** NLP analyzes user feedback and sentiment to assess satisfaction and identify areas for improvement.

- **Adaptive Dialogue Systems:** Conversational agents learn from interactions to provide personalized recommendations and explanations.

These capabilities make personalized network management accessible and user-friendly.

Multi-agent Systems for Collaborative Personalization

In large-scale or multi-domain networks, distributed AI agents collaborate to provide personalized management:

- **Agent Specialization:** Different agents may focus on specific network segments, functions, or user groups, leveraging local knowledge.

- **Coordination and Negotiation:** Agents communicate to resolve conflicts, share insights, and optimize global network objectives.

- **Scalability and Fault Tolerance:** Distributed agents reduce single points of failure and distribute computational load.

- **Use Cases:** Multi-agent systems manage personalized QoS enforcement across enterprise campuses, coordinate security policies in multi-tenant clouds, and optimize resource allocation in 5G networks.

This collaborative approach enhances robustness and scalability.

Explainable AI for Trust and Transparency

As AI agents autonomously manage networks, transparency is essential to build trust and facilitate human oversight:

- **Explainability Methods:** Techniques like SHAP (SHapley Additive exPlanations), LIME (Local Interpretable Model-agnostic Explanations), and attention visualization reveal which features influenced AI decisions.

- **User and Operator Trust:** Clear explanations help users and network operators understand why certain policies or actions were taken.

- **Compliance and Auditing:** Explainable AI supports regulatory compliance by providing audit trails and justifications.

- **Interactive Interfaces:** Visualization tools allow exploration of AI decision processes.

Explainability bridges the gap between complex AI models and human users, ensuring accountability.

User Experience and Satisfaction

To delve deeper, we first need to understand how user experience is measured and evaluated.

Measuring User Experience (UX)

Evaluating the effectiveness of personalized network management requires comprehensive measurement of user experience:

- **Quality of Experience (QoE) Metrics:** Objective metrics such as video startup delay, buffering ratio, call drop rate, page load time, and throughput directly impact perceived user satisfaction.

- **Subjective Feedback:** User surveys, ratings, and sentiment analysis capture qualitative aspects of experience.

- **Network Key Performance Indicators (KPIs):** Metrics like latency, jitter, packet loss, and availability are correlated with QoE to diagnose issues.

- **Passive and Active Monitoring:** Network probes and synthetic transactions provide real-time performance data.

- **AI-Driven Correlation:** Machine learning models analyze multidimensional data to identify root causes of poor UX and predict satisfaction levels.

Holistic UX measurement enables continuous improvement of personalized management strategies.

Impact of Personalization on User Satisfaction

Personalized network management improves user satisfaction by

- **Reducing Latency and Jitter:** Tailored routing and bandwidth allocation minimize delays, crucial for interactive applications like gaming and video calls.

- **Enhancing Reliability:** Adaptive policies prevent service disruptions and maintain consistent performance.

- **Improving Security:** Customized security measures protect users without imposing unnecessary restrictions, reducing false alarms and interruptions.

- **Increasing Transparency and Control:** Explainable AI and user interfaces empower users to understand and influence their network experience.

- **Supporting Diverse Needs:** Personalization accommodates varying user priorities, whether speed, cost, security, or convenience.

Higher satisfaction leads to increased loyalty, reduced churn, and positive word-of-mouth.

Challenges in Optimizing UX

Despite its benefits, optimizing UX through personalization faces challenges:

- **Balancing Conflicting Objectives:** Enhancing one user's experience may degrade others'; AI must ensure fairness and equitable resource distribution.

- **Privacy Preservation:** Collecting and analyzing user data must comply with privacy laws and respect user consent.

- **Latency in Adaptation:** Personalization adjustments must occur with minimal delay to avoid disrupting ongoing sessions.

- **User Autonomy vs. Automation:** Providing users with control options while leveraging AI-driven automation requires careful interface and policy design.

Addressing these challenges is key to sustainable personalization.

Future Directions in User-Centric Networking

Looking forward, several trends will shape the future of personalized network management:

- **Context-Aware Personalization:** Incorporating richer context such as emotional state, social environment, and device health to refine personalization

- **Cross-Domain Integration:** Seamlessly integrating personalization across networks, cloud services, edge computing, and applications for end-to-end optimization

- **Federated Learning:** Collaborative AI model training across multiple entities without sharing raw data, preserving privacy while enhancing model accuracy

- **Adaptive User Interfaces:** Interfaces that evolve based on user expertise, preferences, and feedback to improve engagement and control

- **Ethical AI Practices:** Ensuring fairness, transparency, and accountability in personalization algorithms

These advances will deepen the synergy between users and networks, creating more responsive and human-centric digital environments.

Personalized Network Management with AI Agents: Detailed Conceptual Diagram

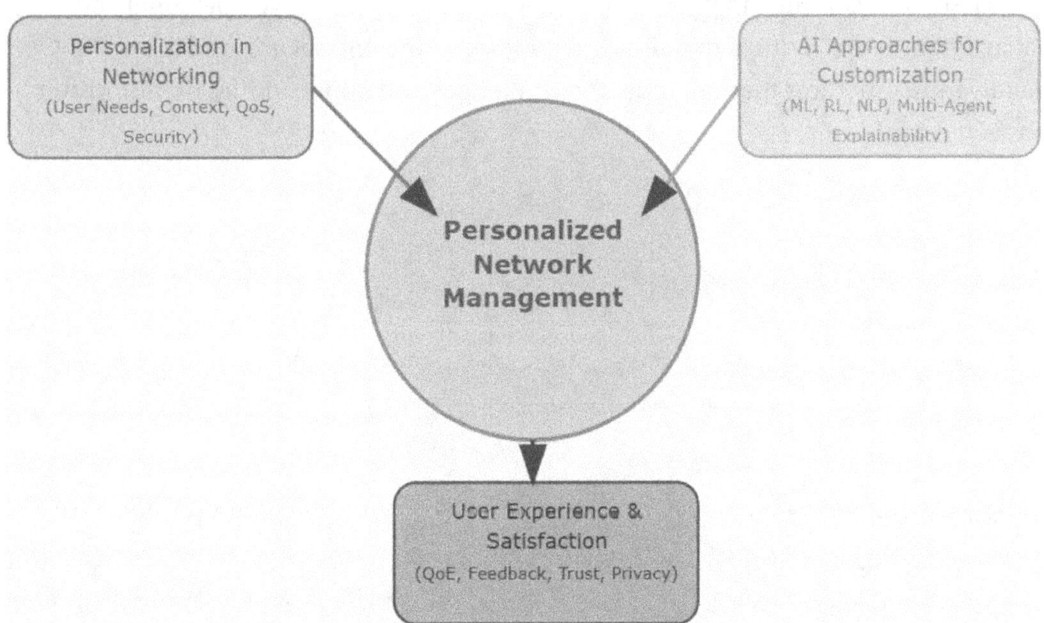

Figure 17-1. *Detailed Conceptual Diagram of Personalized Network Management with AI Agents*

This figure highlights the interplay between personalization in networking and AI approaches feeding into the core concept, which leads to enhanced user experience and satisfaction. Key components and subdomains are annotated for clarity.

Conclusion

Personalized network management is reshaping the future of networking by shifting from rigid, uniform policies to flexible, user-centric configurations that respond dynamically to individual needs and contexts. AI agents are indispensable enablers of this transformation, automating complex tasks, learning from vast data, and continuously adapting network behavior to optimize performance, security, and user satisfaction.

Through sophisticated techniques such as user profiling, reinforcement learning, natural language processing, multi-agent collaboration, and explainable AI, networks can deliver tailored services that meet diverse and evolving demands. Measuring and enhancing user experience remains a central goal, ensuring that personalization translates into tangible benefits.

As networks continue to grow in scale and complexity, AI-driven personalized management will be critical to building responsive, efficient, secure, and user-friendly infrastructures that can meet the challenges and opportunities of tomorrow's connected world.

CHAPTER 18

AI-Driven Energy Efficiency in SDN

In today's hyper-connected world, Software-Defined Networking (SDN) stands as a transformative technology, revolutionizing how networks are built, managed, and optimized. Its inherent programmability and centralized control empower network operators with unprecedented flexibility to adapt to rapidly changing traffic demands and complex service requirements. However, this flexibility and scalability come at a cost: increased energy consumption.

As data traffic continues to grow exponentially, fueled by cloud computing, Internet of Things (IoT), 5G, and multimedia applications, the energy footprint of network infrastructures has become a pressing concern. Network operators face mounting operational expenses and environmental responsibilities, making energy efficiency a critical priority.

> *Energy efficiency is not just a technical goal but a commitment to a sustainable future—where intelligent networks power progress without compromising our planet.*

Artificial intelligence (AI), with its ability to analyze vast datasets, learn patterns, and make autonomous decisions, offers a promising avenue to tackle these challenges. By integrating AI into SDN, networks can dynamically optimize energy consumption while ensuring high performance and reliability.

This chapter explores the multifaceted energy consumption challenges inherent in SDN environments, presents AI-driven solutions that address these issues, and discusses emerging trends that promise to shape the future of sustainable network management.

© Het Mehta 2026
H. Mehta, *AI Agents for Secure and Software-Defined Networking*,
https://doi.org/10.1007/979-8-8688-2358-9_18

Energy Consumption Challenges

The journey toward energy-efficient SDN begins with a clear understanding of the scope and nature of energy consumption within modern networks. The rapid expansion of digital services has led to a corresponding surge in the energy demands of network infrastructures.

The Growing Energy Footprint of Modern Networks

Networks today are sprawling ecosystems comprising data centers, edge devices, core and access routers, and countless communication links. Each component contributes to the overall energy profile in different ways. The explosion of bandwidth-hungry applications like video streaming, virtual reality, and real-time analytics has pushed network utilization to new heights. Yet, despite this growth, many network elements continue to consume significant power even during periods of low activity.

For instance, network devices such as switches and routers often remain fully powered regardless of traffic load, leading to inefficiencies. Moreover, to guarantee reliability and fault tolerance, networks maintain redundant paths and reserve capacity, resulting in over-provisioning. This redundancy, while essential for resilience, compounds energy consumption.

In addition, the migration toward virtualized network functions (VNFs) and cloud-native architectures introduces further energy demands on compute and cooling systems. Collectively, these factors contribute to networks accounting for a substantial fraction of global ICT energy use, raising both economic and environmental concerns.

Energy Consumption Across SDN Components

Delving deeper, it is important to recognize how energy is distributed across the various components of an SDN architecture. The SDN controller, which orchestrates the network's behavior, runs on servers that require continuous power to maintain responsiveness and handle control plane functions. The data plane devices—switches and routers—consume energy based on their forwarding activity, port usage, and hardware design characteristics.

Communication links, whether optical or electrical, also draw power that scales with traffic load and link distance. Virtualized network functions add another layer, running on shared compute infrastructure that must be efficiently managed to avoid unnecessary energy waste. Finally, monitoring and telemetry systems, essential for gathering data to inform AI algorithms, contribute to the overall energy footprint.

Understanding these components and their energy profiles lays the groundwork for targeted optimization strategies.

Challenges to Energy Efficiency in SDN

Despite the clear need for energy reduction, several challenges complicate the path forward. One of the primary obstacles is the trade-off between energy savings and network performance. Aggressive power-down strategies may reduce energy consumption but risk increasing latency, packet loss, or reducing throughput, which can degrade user experience.

Moreover, the dynamic and heterogeneous nature of modern networks—with fluctuating traffic patterns, varying device capabilities, and evolving topologies—requires energy management solutions that are adaptive and scalable. Many network devices lack fine-grained power control, limiting the options for energy savings at the hardware level.

Finally, any energy optimization must preserve service quality and reliability, ensuring that users do not experience disruptions or degraded performance. These complexities underscore the need for intelligent, data-driven approaches that can balance competing demands effectively.

AI Solutions for Energy Efficiency

Addressing these challenges calls for innovative solutions, and artificial intelligence stands out as a powerful enabler. By leveraging AI's capabilities in data analysis, pattern recognition, and autonomous decision-making, SDN can achieve dynamic, fine-grained energy optimization that adapts to real-time conditions.

Traffic Prediction and Load Forecasting

A fundamental step toward energy-efficient management is anticipating network demand. AI-powered traffic prediction models analyze historical data to forecast future loads, identifying patterns such as daily peaks, seasonal fluctuations, and unexpected bursts.

Time series forecasting techniques, including ARIMA and advanced neural networks like Long Short-Term Memory (LSTM), excel at capturing complex temporal dependencies. Complementing these, machine learning classifiers can categorize traffic by type—distinguishing, for example, between latency-sensitive voice calls and bulk data transfers—allowing for differentiated energy policies.

With accurate predictions, network controllers can proactively adjust resource allocation, powering down underutilized devices during predicted low-traffic intervals. This foresight prevents wasted energy without compromising readiness for sudden demand spikes.

Dynamic Resource Management

Building on prediction capabilities, AI facilitates real-time resource management to optimize energy consumption dynamically. Adaptive Link Rate (ALR) techniques enable the modulation of link speeds based on current traffic, reducing power draw during off-peak times.

Machine learning algorithms identify idle or lightly loaded switches and ports suitable for sleep modes, safely transitioning devices to low-power states and reactivating them seamlessly when needed. Additionally, AI-driven load balancing consolidates traffic flows onto fewer active devices, maximizing utilization and enabling others to be powered down.

Virtualized network function placement and migration further enhance efficiency by optimizing server workloads, minimizing idle compute resources. These dynamic adjustments collectively yield substantial energy savings while maintaining network performance.

Reinforcement Learning for Energy-Performance Trade-offs

Reinforcement learning (RL) offers a sophisticated approach to balancing energy efficiency with network performance. Through iterative interaction with the network environment, RL agents learn policies that maximize a reward function designed to penalize excessive energy use and performance degradation.

This method enables continuous adaptation to changing network conditions without explicit reprogramming. Multi-agent RL frameworks extend this concept by distributing control across network segments, coordinating actions to optimize energy use globally.

Experimental implementations have demonstrated that RL can achieve energy reductions up to 30%, with negligible impact on latency and throughput, showcasing its practical potential for real-world SDN deployments.

AI-Enhanced Network Topology Optimization

AI also plays a crucial role in optimizing the network's physical and logical topology for energy savings. By analyzing traffic patterns and device utilization, AI models can identify redundant links and devices that can be temporarily disabled without affecting connectivity or resilience.

Graph Neural Networks (GNNs), capable of modeling complex network relationships, assist in discovering optimal configurations that traditional heuristics might overlook. Dynamic routing adjustments concentrate traffic on fewer active paths, enabling the network to operate with a reduced energy footprint while maintaining robustness.

Predictive Maintenance and Fault Management

Energy efficiency is closely linked to network reliability. Malfunctioning or degraded hardware often operates inefficiently, consuming more power than necessary. AI-driven predictive maintenance uses anomaly detection techniques to identify early signs of failure, allowing for proactive repairs.

By scheduling maintenance before faults occur, networks avoid energy waste and unexpected outages. Furthermore, energy management policies can adapt based on device health, preventing overloading and inefficiencies.

AI-Driven Energy-Aware Network Slicing

In multi-tenant SDN and 5G environments, network slicing enables the creation of multiple virtual networks on shared physical infrastructure. AI enhances energy efficiency by orchestrating slice resources with energy-awareness in mind.

Dynamic scaling of slices based on traffic and energy goals ensures that resources are not wasted during low demand. This fine-grained control balances tenant requirements with sustainability objectives, promoting efficient infrastructure use.

Future Trends in Energy Management

Looking ahead, several emerging trends promise to further advance AI-driven energy efficiency in SDN, integrating new technologies and addressing evolving challenges.

Integration of AI with Green Networking Standards

The development of green networking standards provides a framework for sustainable network operation. AI will be deeply integrated into these standards, enabling automated energy optimization with standardized metrics and interfaces.

This integration will facilitate interoperability among diverse vendors and platforms, fostering widespread adoption of AI energy management solutions aligned with environmental regulations and corporate sustainability goals.

Edge and Fog Computing for Distributed Energy Optimization

The rise of edge and fog computing decentralizes network functions closer to the user, distributing energy management responsibilities. Localized AI agents at edge nodes will optimize energy consumption based on real-time, localized traffic and device conditions.

Collaboration between edge and cloud AI systems will enable holistic energy optimization across the entire network stack, reducing latency and improving responsiveness while maintaining efficiency.

Federated Learning for Privacy-Preserving Optimization

Privacy concerns often limit data sharing across network domains. Federated learning offers a solution by enabling collaborative AI model training without exchanging raw data.

This approach allows multiple operators to jointly improve energy management models while preserving data confidentiality, enhancing model accuracy and generalization across diverse network environments.

Explainable AI for Transparency and Trust

As AI assumes greater control over energy management, transparency becomes critical. Explainable AI techniques provide insights into AI decision-making processes, fostering trust among network operators.

Interpretable models support human-in-the-loop oversight, enabling operators to understand, validate, and intervene in AI-driven energy policies, ensuring compliance with regulations and organizational objectives.

AI-Driven Renewable Energy Integration

Future networks will increasingly incorporate renewable energy sources such as solar and wind. AI will optimize the use of harvested energy by scheduling network workloads to align with renewable availability.

Integration with smart grids will allow networks to adjust consumption dynamically based on grid conditions and energy prices, promoting sustainability and cost savings.

Conclusion

Energy efficiency in SDN is a complex but essential goal, driven by the twin imperatives of cost reduction and environmental stewardship. AI offers transformative capabilities to meet this challenge, enabling networks to predict demand, adapt resources dynamically, learn optimal policies, optimize topology, and maintain health proactively.

As AI technologies evolve and integrate with emerging standards, edge computing, privacy-preserving techniques, explainability, and renewable energy systems, the vision of sustainable, intelligent networks comes into sharper focus. Embracing AI-driven energy efficiency is not only a technological advancement but a vital commitment to a greener digital future.

CHAPTER 19

AI-Powered Collaborative Security for Multi-domain SDN

The advent of Software-Defined Networking (SDN) has revolutionized network management by decoupling control and data planes, enabling programmability and flexibility. However, as SDN deployments expand across multiple administrative domains—each with distinct policies and infrastructures—the security landscape grows increasingly complex. Multi-domain SDN environments interconnect independently managed networks, creating rich ecosystems that, while powerful, present unique security challenges.

Traditional security models, designed for isolated or single-domain networks, fall short in addressing the intricacies of multi-domain collaboration. Isolated defenses cannot effectively detect or mitigate threats that traverse domain boundaries, nor can they coordinate responses swiftly enough to prevent widespread damage.

In an interconnected world, security is no longer confined within borders—collaboration and intelligence are the keys to safeguarding multi-domain networks.

Artificial intelligence (AI), with its advanced data analytics, anomaly detection, and autonomous decision-making capabilities, emerges as an essential tool for collaborative security. By enabling domains to share threat intelligence, coordinate defense strategies, and automate incident responses, AI enhances the resilience of multi-domain SDN infrastructures.

This chapter explores the challenges inherent in securing multi-domain SDN, details AI-driven collaborative security approaches, and presents real-world case studies demonstrating their effectiveness.

© Het Mehta 2026
H. Mehta, *AI Agents for Secure and Software-Defined Networking*,
https://doi.org/10.1007/979-8-8688-2358-9_19

Multi-domain Security Challenges

Securing a multi-domain SDN is inherently more complex than securing a single domain due to several interrelated factors. Understanding these challenges is critical for designing effective AI-powered collaborative security solutions.

Diverse Administrative and Policy Boundaries

Each domain in a multi-domain SDN is governed by its own administrative authority, which enforces specific security policies, access controls, and governance models. These differences create barriers to seamless security cooperation. Conflicting policies or lack of standardized protocols impede unified threat detection and coordinated mitigation.

Increased Attack Surface and Threat Vectors

Interconnecting multiple domains expands the attack surface significantly. Attackers can exploit vulnerabilities in one domain to gain access to others, enabling lateral movement and multi-stage attacks. Cross-domain data exfiltration, supply chain attacks, and coordinated distributed denial of service (DDoS) campaigns are amplified in such environments.

Limited Visibility and Information Sharing

Effective security depends on comprehensive visibility. However, domains often hesitate to share detailed security telemetry due to privacy concerns, competitive interests, or regulatory constraints. This fragmentation limits the ability to detect cross-domain threats and hampers collective defense.

Heterogeneous Technologies and Protocols

Multi-domain SDNs typically deploy diverse controllers, devices, and security tools. This heterogeneity complicates interoperability, making it challenging to enforce consistent security policies or deploy unified monitoring and response mechanisms.

Real-Time Threat Detection and Response

The dynamic and distributed nature of multi-domain SDNs demands rapid threat detection and response. Manual coordination is slow and error-prone, necessitating automated, intelligent systems capable of real-time collaboration.

Table 19-1. *Summary of Multi-domain SDN Security Challenges*

Challenge	Description	Impact on Security
Diverse Administrative Policies	Different governance and security policies per domain	Hinders unified threat detection and response
Expanded Attack Surface	More entry points and vectors across interconnected domains	Increases risk of lateral attacks and breaches
Limited Visibility and Sharing	Reluctance to share sensitive security data	Reduces situational awareness and early warnings
Heterogeneous Technologies	Varied controllers, devices, and protocols	Complicates interoperability and policy enforcement
Need for Real-Time Response	Fast-moving threats require immediate coordinated action	Manual processes are insufficient and slow

Understanding these challenges highlights why traditional security approaches are inadequate and underscores the need for AI-driven collaborative frameworks.

Collaborative Approaches with AI

Artificial intelligence provides the foundation for overcoming multi-domain security challenges through enhanced data analysis, automated coordination, and adaptive defense mechanisms.

AI-Enabled Threat Intelligence Sharing

AI systems aggregate security data from multiple domains, including logs, alerts, and telemetry, to generate comprehensive threat intelligence. Machine learning models analyze this data to identify emerging patterns and classify threats with high accuracy.

To address privacy and regulatory concerns, techniques such as federated learning allow AI models to be trained collaboratively without sharing raw data, preserving confidentiality while improving detection capabilities.

Distributed Anomaly Detection Using Machine Learning

Collaborative anomaly detection frameworks use AI models trained on diverse, multi-domain datasets to spot unusual activities indicative of attacks. These models can detect subtle deviations from normal behavior, such as lateral movement or zero-day exploits, that might evade traditional signature-based systems.

By sharing anomaly alerts across domains, the system enables early warning and coordinated defense before attacks escalate.

Automated Incident Response and Orchestration

AI-driven automation orchestrates cross-domain incident responses by coordinating actions like dynamic traffic filtering, quarantining compromised nodes, and updating security policies in real time. Reinforcement learning agents continuously optimize these actions based on feedback, balancing security effectiveness with minimizing operational disruptions.

This automation accelerates response times and reduces human error in complex multi-domain scenarios.

Trust Management and Access Control

AI facilitates dynamic trust evaluation among domains by analyzing behavior, historical interactions, and threat intelligence. This enables adaptive access control policies that evolve in response to changing risk levels, allowing secure collaboration without sacrificing agility.

Blockchain and AI for Secure Collaboration

Integrating blockchain with AI enhances trustworthiness and transparency in multi-domain security. Blockchain's immutable ledger records security events and policy changes, ensuring accountability and auditability. AI analyzes these records to detect policy violations and automate compliance enforcement.

This combination strengthens cooperation among domains by providing a secure, verifiable foundation for collaboration.

AI-Powered Collaborative Security Architecture

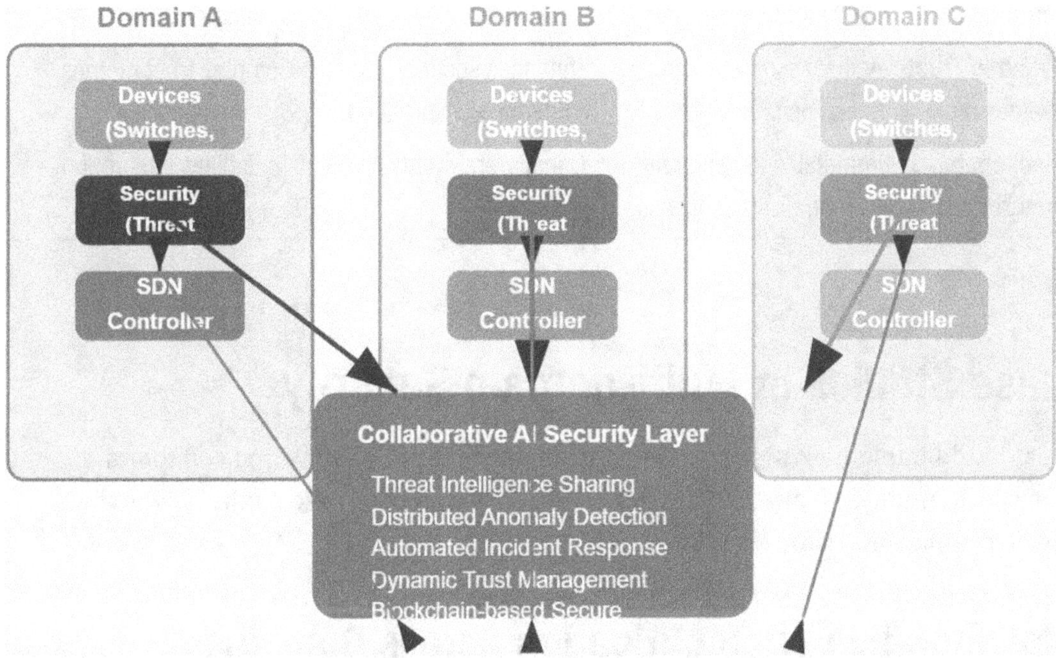

Figure 19-1. *AI-Powered Collaborative Security Architecture for Multi-domain SDN*

This architecture illustrates how local AI security modules within each domain communicate with a collaborative AI security layer. The collaborative layer aggregates threat intelligence, performs distributed anomaly detection, orchestrates incident response, and manages trust and blockchain-enabled secure collaboration.

Table 19-2. *Summary of AI Techniques in Multi-domain Security*

AI Technique	Description	Benefits	Challenges
Federated Learning	Collaborative model training without data sharing	Preserves privacy while improving detection	Communication overhead, model convergence
Anomaly Detection	Identifies deviations from normal behavior	Early detection of unknown threats	False positives, dataset diversity
Reinforcement Learning	Optimizes automated incident response	Fast, adaptive mitigation	Requires extensive training data
Dynamic Trust Evaluation	Adjusts access control based on behavior	Enhances security without hindering collaboration	Complex trust metrics, dynamic policies
Blockchain Integration	Immutable logging and decentralized consensus	Transparency and auditability	Scalability, integration complexity

Case Studies in Multi-domain Security

Real-world implementations validate the effectiveness of AI-powered collaborative security in multi-domain SDN environments. The following case studies highlight diverse applications and outcomes.

Collaborative DDoS Mitigation Across Domains

A consortium of Internet Service Providers (ISPs) deployed an AI-based system to collaboratively detect and mitigate large-scale DDoS attacks. Each domain's AI module monitored local traffic, sharing anomaly alerts with a centralized collaborative AI engine.

Machine learning classifiers identified attack signatures and coordinated traffic filtering rules across domains, effectively throttling malicious flows. The system reduced the attack impact by approximately 40% in terms of traffic volume reaching critical infrastructure and duration of service degradation compared to isolated defenses, demonstrating the power of collaboration.

Federated Learning for Cross-Domain Intrusion Detection

Multiple enterprise networks participated in a federated learning initiative to develop a shared intrusion detection model. This approach enabled training on distributed datasets without exposing sensitive security logs.

The resulting AI model improved detection rates of novel and zero-day threats by 25%, enhancing security posture while maintaining strict data privacy. This case illustrates how AI can bridge collaboration and confidentiality.

AI-Orchestrated Incident Response in Multi-operator Networks

In a multi-operator SDN deployment supporting smart city infrastructure, reinforcement learning agents coordinated automated responses to security incidents. Upon detecting anomalies, the system dynamically adjusted firewall policies, rerouted traffic, and quarantined affected segments.

This orchestration reduced incident resolution time by 35% and minimized service disruptions, highlighting AI's role in accelerating and optimizing complex responses.

Blockchain-Backed Security Policy Enforcement

A group of financial institutions implemented a blockchain-based framework to record security policy changes across their SDN domains. AI monitored compliance by analyzing blockchain records for policy violations or suspicious activity.

This transparent, tamper-proof system enhanced trust among participants and ensured accountability, reducing the risk of insider threats and misconfigurations.

Conclusion

The security of multi-domain SDN environments is challenged by diversity, scale, and evolving threats. Traditional siloed defenses cannot keep pace with these complexities. AI-powered collaborative security offers a robust path forward by enabling intelligent threat sharing, distributed detection, automated response, and adaptive trust management.

The integration of AI with emerging technologies such as blockchain further strengthens the foundation for secure, transparent collaboration. Real-world case studies demonstrate tangible improvements in threat mitigation, detection accuracy, and operational efficiency.

As multi-domain SDNs continue to grow in importance, embracing AI-driven collaborative security will be essential for safeguarding critical infrastructure and maintaining trust in interconnected digital ecosystems.

CHAPTER 20

AI in Programmable Optical Networks with SDN

The explosion of data traffic fueled by cloud computing, Internet of Things (IoT), 5G, and emerging applications such as virtual reality and autonomous vehicles demands communication networks that are not only high-capacity but also highly flexible and intelligent. Traditional optical networks, while offering immense bandwidth, are often rigid and manually configured, limiting their ability to adapt swiftly to dynamic traffic patterns and service requirements.

Programmable optical networks (PONs), empowered by Software-Defined Networking (SDN), represent a paradigm shift in optical communications. By decoupling the control plane from the data plane, PONs enable centralized, software-driven control of optical network elements, allowing dynamic configuration of wavelengths, routes, and resources. This programmability brings agility, scalability, and operational efficiency that are essential for modern networks.

> *Harnessing AI to unlock the full potential of programmable optical networks integrated with SDN, paving the way for smarter, faster, and more resilient communication infrastructures.*

Artificial intelligence (AI) complements this transformation by introducing intelligent automation, predictive analytics, and adaptive decision-making capabilities. AI algorithms can analyze vast amounts of network data, identify patterns, predict failures, and optimize network parameters in real-time. When integrated with SDN-controlled programmable optical networks, AI can significantly improve network performance, reliability, and security.

© Het Mehta 2026
H. Mehta, *AI Agents for Secure and Software-Defined Networking,*
https://doi.org/10.1007/979-8-8688-2358-9_20

This chapter delves into the fundamentals of programmable optical networks, explores the expanding role of AI in optical networking, and discusses the technical and operational challenges faced when applying AI in this domain. The chapter also highlights future directions and best practices for integrating AI in programmable optical networks with SDN.

Overview of Programmable Optical Networks

To grasp the concept of programmable optical networks, we first need to define them and examine their architecture.

Definition and Architecture

Programmable optical networks are advanced optical communication systems designed to be dynamically controlled and reconfigured via software interfaces, enabling network operators to adapt resources according to real-time demands. Unlike traditional static optical networks, which rely on fixed hardware configurations and manual interventions, PONs employ SDN principles to separate the control plane from the data plane.

This separation allows centralized SDN controllers to manage optical devices such as Reconfigurable Optical Add-Drop Multiplexers (ROADMs), optical amplifiers, and transponders through standardized southbound interfaces. Common protocols used include NETCONF/YANG for device configuration and OpenFlow extensions tailored for optical networks.

Architecture Components

- **Optical Data Plane:** Includes fiber links, ROADMs, optical amplifiers, transponders, and wavelength selective switches. This plane handles the actual transmission of optical signals.

- **SDN Control Plane:** Centralized controllers that orchestrate the optical devices, managing routing, wavelength assignment, and resource allocation dynamically.

- **Network Management and Orchestration:** Higher-level systems that provide policy enforcement, monitoring, fault management, and service provisioning.

Key Features and Benefits

- **Dynamic Resource Allocation:** Enables on-demand provisioning and adjustment of wavelengths, bandwidth, and routes, improving responsiveness to traffic fluctuations.

- **Flexibility and Programmability:** Allows rapid adaptation to changing service requirements, supporting multi-tenancy and network slicing.

- **Scalability:** Facilitates the growth of network size and complexity without proportional increases in operational overhead.

- **Improved Utilization:** Maximizes the efficiency of optical spectrum and hardware resources, reducing waste.

- **Simplified Operations:** Centralized control reduces manual configuration errors and accelerates deployment of new services.

- **Interoperability:** Use of standardized interfaces promotes integration across multi-vendor environments.

Use Cases

- **Internet Service Providers (ISPs):** Backbone networks requiring high-capacity, flexible routing to support diverse customer demands

- **Data Center Interconnects:** Enabling low-latency, high-throughput connections between geographically distributed data centers

- **5G Transport Networks:** Supporting stringent latency, bandwidth, and reliability requirements for mobile backhaul and fronthaul

- **Disaster Recovery and Resilience:** Fast reconfiguration in response to failures or attacks, ensuring service continuity

AI Applications in Optical Networking

Artificial intelligence is increasingly being integrated into programmable optical networks to enhance automation, optimize performance, and improve reliability. The following subsections explore the primary AI applications in this domain.

Traffic Prediction and Demand Forecasting

Accurate traffic prediction is critical for proactive resource management in optical networks. AI models, particularly supervised machine learning algorithms such as regression, time series forecasting (e.g., LSTM networks), and ensemble methods, analyze historical and real-time traffic data to forecast future demand patterns.

By anticipating traffic surges or drops, network controllers can preemptively allocate wavelengths and adjust routing paths, thereby minimizing congestion and improving Quality of Service (QoS). This proactive approach reduces latency and packet loss, enhancing user experience.

Fault Detection and Localization

Optical networks are susceptible to various faults, including fiber cuts, amplifier failures, and signal degradation. Traditional monitoring methods often detect faults only after service disruption occurs.

AI techniques, including unsupervised learning and deep learning, can analyze optical signal parameters such as Optical Signal-to-Noise Ratio (OSNR), Bit Error Rate (BER), and power levels to detect anomalies early. Convolutional Neural Networks (CNNs) and Autoencoders are used for anomaly detection without requiring labeled fault data.

Furthermore, AI can assist in precise fault localization by correlating multi-point measurements and historical fault patterns, significantly reducing mean time to repair (MTTR) and improving network availability.

Dynamic Network Optimization

Reinforcement learning (RL) algorithms enable autonomous decision-making for complex network optimization tasks. RL agents interact with the network environment to learn optimal policies for routing, wavelength assignment, and power control.

For example, RL can dynamically select routes that balance load and minimize latency while adapting to network failures or changing traffic conditions. This continuous learning approach improves network efficiency and robustness over time.

Adaptive Modulation and Coding

Optical channels exhibit varying conditions due to physical impairments and environmental factors. AI algorithms can dynamically adjust modulation formats (e.g., QPSK, 16-QAM) and Forward Error Correction (FEC) coding schemes based on real-time channel quality assessments.

This adaptive approach enhances spectral efficiency and link robustness, enabling higher data rates without compromising signal integrity.

Security Enhancements

Optical networks face security threats such as jamming, eavesdropping, and physical tampering. AI can detect unusual patterns in optical signals indicative of attacks by analyzing parameters like sudden power drops or unexpected noise.

Integrating AI with SDN controllers allows automated mitigation strategies, such as rerouting traffic or isolating affected segments, thus strengthening network security.

Table 20-1. *AI Techniques and Their Applications in Programmable Optical Networks*

AI Technique	Application Area	Benefits
Supervised Learning	Traffic prediction	Accurate demand forecasting, efficient resource planning
Unsupervised Learning	Fault detection	Early anomaly detection without labeled data
Reinforcement Learning	Dynamic routing and optimization	Autonomous decision-making, improved QoS
Deep Learning	Fault localization	Precise identification of fault locations
Evolutionary Algorithms	Modulation and coding adaptation	Enhanced spectral efficiency, adaptive performance

Challenges in Optical Networks

Despite the promising benefits of AI integration, several technical and operational challenges must be addressed.

Data Collection and Quality

High-quality, comprehensive datasets are essential for training effective AI models. However, collecting such data in optical networks is challenging due to

- Limited instrumentation and monitoring capabilities in optical hardware

- Proprietary interfaces and vendor-specific protocols restricting data access

- Noise and variability in optical measurements affecting data reliability

Insufficient or poor-quality data can lead to inaccurate models and suboptimal decisions.

Real-Time Processing Requirements

Optical networks operate at extremely high speeds, often in the order of terabits per second. AI algorithms must process large volumes of data and make decisions with minimal latency to be effective.

Balancing model complexity and computational efficiency is critical. Lightweight, optimized AI models and edge computing approaches are often necessary to meet real-time constraints.

Model Generalization and Adaptability

Optical network environments vary widely due to differences in topology, hardware, and traffic patterns. AI models trained in one network may not perform well in another without adaptation.

Developing models that generalize across diverse conditions or can be efficiently retrained or fine-tuned is a significant challenge.

Integration with Legacy Systems

Many existing optical networks include legacy equipment lacking programmability or standardized interfaces, complicating AI integration.

Bridging the gap between modern SDN-enabled devices and legacy hardware requires hybrid control architectures and careful interoperability planning.

Security and Privacy Concerns

Sharing operational data across domains or operators for collaborative AI training raises privacy and security issues. Sensitive information could be exposed, or adversaries could manipulate AI models.

Secure data sharing protocols, federated learning, and robust AI model validation are needed to mitigate these risks.

Table 20-2. *Key Challenges in Applying AI to Programmable Optical Networks*

Challenge	Description	Impact on AI Deployment
Data Quality and Availability	Limited instrumentation and proprietary data	Hinders accurate model training and validation
Real-Time Constraints	Need for ultra-fast processing and decisions	Limits complexity of AI models used
Model Transferability	Variability across network environments	Requires adaptable and robust AI solutions
Legacy Equipment	Lack of programmability in older devices	Restricts AI-driven automation and control
Security and Privacy	Data sharing risks in collaborative settings	Necessitates secure data handling mechanisms

Future Directions and Best Practices

To fully realize the benefits of AI in programmable optical networks, ongoing research and development focus on

- **Standardization:** Developing open, vendor-neutral data models and interfaces to facilitate AI integration

- **Hybrid AI Architectures:** Combining centralized and distributed AI models for scalability and real-time responsiveness

- **Explainable AI:** Enhancing transparency and trust in AI decisions to support operator acceptance

- **Federated Learning:** Enabling collaborative AI model training across multiple operators without sharing raw data

- **Edge Computing:** Deploying AI inference closer to the optical devices to reduce latency

- **Robust Security Measures:** Implementing AI-driven security frameworks that can detect and mitigate novel threats

Network operators are encouraged to adopt incremental AI integration strategies, starting with noncritical applications such as monitoring and analytics, before expanding to automated control and orchestration.

Conclusion

Programmable optical networks integrated with SDN offer a powerful foundation for building intelligent, flexible, and scalable communication infrastructures. AI technologies enhance these networks by enabling predictive analytics, autonomous optimization, fault management, and security.

While challenges remain in data acquisition, real-time processing, and system integration, advances in AI algorithms, computing architectures, and standardization efforts are progressively overcoming these barriers.

The synergy of AI, SDN, and programmable optical networks will be pivotal in supporting the data-intensive, latency-sensitive applications of the future, driving innovation in telecommunications and beyond.

CHAPTER 21

AI-Driven Predictive Analytics for SDN Scalability

Software-Defined Networking (SDN) has fundamentally transformed network architecture by decoupling the control plane from the data plane, enabling centralized network management, programmability, and flexibility. This architectural shift simplifies network operations and accelerates service deployment. However, as networks expand in size, complexity, and traffic volume, scalability challenges emerge that can undermine the benefits of SDN.

The centralized nature of SDN controllers, while powerful, can become a bottleneck when managing large numbers of devices and flows. Network state management, flow table limitations, latency, and multi-controller coordination issues further complicate scalability. Addressing these challenges proactively is critical to maintaining optimal network performance and reliability.

Leveraging AI-powered predictive analytics to anticipate and address scalability challenges in Software-Defined Networking, ensuring robust, efficient, and future-proof network operations.

Artificial intelligence (AI), particularly predictive analytics, offers transformative potential to anticipate network stresses and enable dynamic, data-driven scalability management. By analyzing historical and real-time network data, AI models can forecast traffic trends, controller loads, and potential congestion points, empowering network operators to allocate resources efficiently, balance loads, and scale controllers dynamically before problems arise.

199

© Het Mehta 2026
H. Mehta, *AI Agents for Secure and Software-Defined Networking,*
https://doi.org/10.1007/979-8-8688-2358-9_21

This chapter provides a comprehensive exploration of SDN scalability challenges, introduces a variety of AI-driven predictive analytics techniques suited to these challenges, and presents implementation strategies supported by real-world case studies. Through this, readers will gain a thorough understanding of how AI can enhance SDN scalability and the practical considerations involved.

Scalability Challenges in SDN

Before delving into AI solutions, it is essential to understand the key scalability challenges inherent in SDN architectures. These challenges highlight the critical areas where predictive analytics can provide impactful improvements.

Centralized Controller Bottlenecks

SDN controllers serve as the brains of the network, responsible for decision-making, policy enforcement, and device configuration. As the network grows, the volume of control messages and flow requests directed to controllers increases exponentially. This surge can overwhelm controller CPU and memory resources, leading to processing delays and degraded responsiveness.

Moreover, the limited throughput of controllers can result in delayed flow setups, impacting time-sensitive applications and reducing overall network performance.

Network State Management Complexity

Maintaining a consistent and up-to-date global view of the entire network state is fundamental for effective SDN operation. However, as the number of network devices and links increases, the frequency and volume of state updates sent to the controller also rise. This can saturate communication channels and overwhelm controller processing capabilities.

Inefficient state management may cause stale or inconsistent information, leading to suboptimal routing decisions and potential network instability.

Flow Table Limitations

SDN switches rely on flow tables to make forwarding decisions. These tables have finite capacity, especially in hardware switches using Ternary Content Addressable Memory (TCAM), which is expensive and limited in size.

High volumes of unique flows can exhaust flow table entries, causing table misses and forcing packets to be sent to the controller for handling. This increases controller load and latency, further exacerbating scalability issues.

Latency and Response Time

As networks scale, the latency between switches and controllers can increase. Additionally, the time controllers take to process requests and install flow rules can grow due to higher loads. These delays impact packet forwarding, leading to increased jitter and packet loss, which are detrimental to real-time services such as VoIP, video conferencing, and online gaming.

Multi-controller Coordination

To overcome single-controller limitations, large SDN deployments often utilize distributed or hierarchical controller architectures. While this approach improves scalability, it introduces new challenges related to synchronization of network state, consistency of policies, and coordination overhead between controllers.

Ensuring seamless interoperability among controllers without introducing excessive latency or inconsistencies is a complex task.

Understanding these challenges lays the groundwork for exploring how AI-driven predictive analytics can provide foresight and automation to mitigate scalability constraints effectively.

Predictive Analytics Techniques

Predictive analytics applies AI and machine learning techniques to analyze historical and current data to forecast future network states, enabling proactive network management. Various predictive analytics methods can be tailored to address specific SDN scalability challenges.

Time Series Forecasting

Time series forecasting models analyze sequences of data points collected over time to predict future values. These models are particularly useful for anticipating traffic volume, flow arrival rates, and controller load patterns.

- **ARIMA (AutoRegressive Integrated Moving Average):** A classical statistical method effective for stationary time series with linear trends

- **Seasonal ARIMA (SARIMA):** Extends ARIMA to model seasonal patterns common in network traffic

- **Long Short-Term Memory (LSTM) Networks:** A type of recurrent neural network capable of learning long-term dependencies and nonlinear temporal relationships, making them highly effective for complex traffic patterns

By accurately forecasting traffic surges, these models enable preemptive scaling of controllers and resource allocation.

Classification and Regression Models

Supervised machine learning algorithms can classify network states or predict continuous performance metrics based on input features extracted from network data.

- **Support Vector Machines (SVM):** Effective for binary and multi-class classification tasks such as identifying congestion risk levels

- **Random Forests:** Ensemble decision trees providing robust classification and regression capabilities

- **Gradient Boosting Machines (GBM):** Powerful for handling complex nonlinear relationships in data

These models assist in categorizing network conditions and predicting latency or throughput, informing scalability decisions.

Anomaly Detection

Unsupervised learning methods detect deviations from normal network behavior that may signal impending scalability issues.

- **K-Means Clustering:** Groups similar data points to identify outliers.

- **Isolation Forest:** Efficiently isolates anomalies by partitioning data.

- **Autoencoders:** Neural networks trained to reconstruct normal data patterns; high reconstruction errors indicate anomalies.

Early detection of unusual traffic patterns or controller behavior allows timely intervention to prevent bottlenecks.

Reinforcement Learning

Reinforcement learning (RL) involves agents learning optimal policies through interactions with the environment, receiving rewards or penalties based on actions.

In SDN scalability, RL agents can dynamically allocate resources, balance controller loads, and scale components by learning from network feedback, continuously improving decision-making without explicit programming.

Ensemble Methods

Combining multiple models often yields better predictive performance than individual models alone. Techniques such as bagging, boosting, and stacking reduce variance and bias, enhancing prediction accuracy and robustness.

Table 21-1. *Overview of Predictive Analytics Techniques for SDN Scalability*

Technique	Description	Application Area
Time Series Forecasting	Models temporal data trends	Traffic volume prediction, controller load forecasting
Classification and Regression	Predicts categorical or continuous outcomes	Congestion risk classification, latency prediction
Anomaly Detection	Identifies deviations from normal patterns	Early detection of scalability bottlenecks
Reinforcement Learning	Learns optimal actions through trial and error	Dynamic resource allocation, controller scaling
Ensemble Methods	Combines multiple models for improved accuracy	Robust prediction of network states

The choice of technique depends on the specific scalability challenge, available data, and computational constraints. Often, hybrid approaches combining several methods yield the best results.

Implementation and Results

Having examined the predictive analytics techniques, the next step is to understand their practical implementation within SDN environments and observe their impact through real-world results.

Data Collection and Preprocessing

Effective predictive analytics begins with comprehensive data collection from diverse sources including SDN controllers, switches, and network monitoring tools. Key metrics typically gathered include

- Flow arrival rates and durations

- Controller CPU and memory usage

- Packet latency and jitter

- Error and packet drop rates

- Network topology changes

Preprocessing steps prepare the raw data for modeling:

- **Cleaning:** Removing corrupted or incomplete records

- **Normalization:** Scaling features to a uniform range

- **Feature Extraction:** Deriving relevant attributes such as moving averages, peak traffic times, or flow statistics

- **Handling Missing Data:** Imputation or removal of missing values to maintain dataset integrity

This stage is critical to ensure the quality and reliability of AI model training.

Model Training and Validation

Selected predictive models are trained on historical datasets using supervised or unsupervised learning techniques. Validation methods such as k-fold cross-validation assess model generalizability and prevent overfitting.

Hyperparameter tuning optimizes model configurations, balancing complexity and accuracy. For example, the number of layers and neurons in an LSTM network or the depth of trees in a random forest.

Integration with SDN Controllers

Once trained, AI models are integrated into the SDN control architecture. This integration may involve

- Embedding models within controller software for real-time inference

- Using external analytics platforms interfaced via APIs

- Deploying models at the network edge for low-latency decision-making

Predictions generated by AI models inform controller actions such as

- Instantiating additional controller instances or virtual machines

- Redistributing flow processing loads

- Adjusting flow table management policies

- Preemptively rerouting traffic to avoid congestion

Case Study: Traffic Load Prediction for Controller Scaling

In a large-scale SDN testbed simulating a metropolitan area network, an LSTM-based traffic prediction model was implemented to forecast controller load over 5-minute intervals.

The system used predictions to trigger dynamic scaling, spinning up additional controller instances before anticipated load peaks. This proactive approach prevented controller overloads and maintained low response times.

Results:

- Average controller response time decreased by approximately 30%.

- Flow setup latency improved, reducing packet forwarding delays.

- Overall network throughput increased due to better load distribution.

Case Study: Anomaly Detection for Flow Table Management

An Isolation Forest model was trained on flow statistics to identify anomalous flow patterns indicative of potential flow table exhaustion.

Upon detecting anomalies, the system automatically aggregated similar flow entries and optimized rule installation, reducing flow table misses.

Results:

- Flow table misses decreased by 25%, reducing controller intervention.

- Network stability improved with fewer packet drops.

- Controller load was alleviated, enhancing scalability.

AI-Driven Predictive Analytics Workflow for SDN Scalability

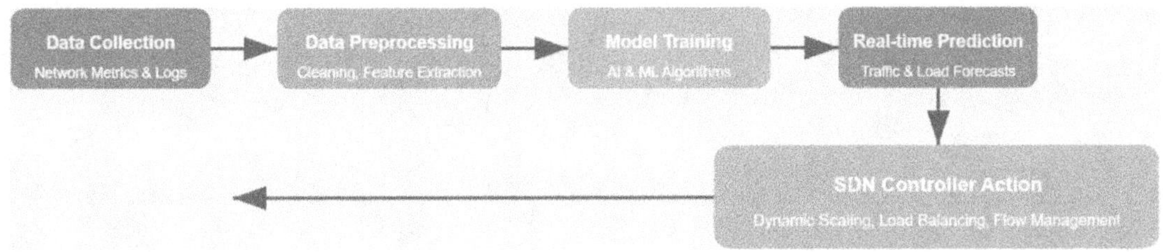

Figure 21-1. *Example Workflow of AI-Driven Predictive Analytics for SDN Scalability*

This workflow illustrates the continuous cycle of data-driven decision-making enabled by AI in SDN scalability management. Data is collected from the network, preprocessed, and used to train predictive models. These models generate real-time forecasts that inform controller actions, which in turn affect network state and generate new data, closing the feedback loop.

Conclusion

AI-driven predictive analytics emerges as a vital enabler for overcoming scalability challenges in SDN environments. By harnessing historical and real-time data, AI models provide foresight into network conditions, facilitating proactive management of controller loads, flow tables, and network state complexity.

The techniques discussed—from time series forecasting and classification to anomaly detection and reinforcement learning—offer versatile tools adaptable to diverse network scenarios. Real-world implementations demonstrate tangible benefits including reduced controller response times, improved flow setup latency, and enhanced network stability.

Nonetheless, successful deployment requires meticulous data management, model selection tailored to network characteristics, and seamless integration with SDN control architectures. Addressing computational efficiency and model adaptability remains an ongoing research focus.

Looking ahead, advancements in federated learning, edge AI, and standardized AI-SDN interfaces will further empower network operators to build scalable, intelligent, and resilient software-defined networks capable of meeting the demands of future digital ecosystems.

CHAPTER 22

Democratizing AI-Driven SDN Management with Open Source

The convergence of Software-Defined Networking (SDN) and Artificial Intelligence (AI) marks a transformative era in network management. SDN's programmability and centralized control enable unprecedented flexibility, while AI introduces intelligent automation, predictive analytics, and adaptive decision-making. Together, they promise networks that are more scalable, resilient, and self-optimizing.

However, despite these advances, the adoption of AI-driven SDN management often remains confined to large enterprises and service providers with significant resources. Proprietary solutions tend to be costly, complex, and closed, limiting accessibility, innovation, and customization. This creates a barrier for smaller organizations, academic researchers, and startups seeking to leverage AI capabilities in network management.

Exploring how open source initiatives empower widespread adoption and innovation in AI-driven Software-Defined Networking management, fostering collaboration, transparency, and rapid evolution.

Open source software and community-driven projects are pivotal in breaking down these barriers. By providing freely accessible, modifiable, and extensible tools, open source democratizes AI-driven SDN management. It fosters collaboration across industry, academia, and hobbyists, accelerating innovation and enabling rapid experimentation and deployment.

© Het Mehta 2026
H. Mehta, *AI Agents for Secure and Software-Defined Networking,*
https://doi.org/10.1007/979-8-8688-2358-9_22

This chapter explores the multifaceted role of open source in networking, focusing on how it supports and accelerates AI integration into SDN management. We examine prominent open source projects, frameworks, and communities driving this evolution. Finally, we look ahead to future trends that will shape the open source AI-SDN landscape, promoting inclusivity, interoperability, and intelligent automation.

The Role of Open Source in Networking

Open source has been a cornerstone in reshaping modern networking. Its principles of transparency, collaboration, and shared ownership align naturally with the goals of SDN and AI-driven management.

Historical Context and Impact

Historically, networking was dominated by proprietary hardware and software, leading to vendor lock-in, high costs, and limited innovation. The rise of open source networking initiatives began to challenge this model, introducing modular software platforms that decouple hardware from control logic.

Projects like OpenFlow, OpenDaylight, and ONOS have been instrumental in pioneering open SDN controllers, enabling programmable networks that can be customized to specific needs. This shift has lowered barriers for experimentation and accelerated the pace of innovation.

Accelerating Innovation and Collaboration

Open source projects create a collaborative environment where diverse stakeholders—including network operators, hardware vendors, software developers, and researchers—contribute to a common codebase. This collective effort accelerates feature development, improves software quality, and expands use cases.

For AI-driven SDN management, open source communities provide shared datasets, model implementations, and integration tools that reduce duplication of effort and foster rapid progress.

Reducing Costs and Avoiding Vendor Lock-In

By eliminating licensing fees and proprietary restrictions, open source solutions significantly reduce operational expenditures. Organizations gain the freedom to customize and extend software to meet evolving requirements without being tied to a single vendor's road map or pricing.

This flexibility is particularly beneficial for AI integration, where experimentation with different algorithms, data sources, and deployment models is essential.

Enhancing Transparency and Security

Access to source code enables thorough security audits, vulnerability assessments, and performance evaluations. Transparency builds trust, especially in critical infrastructure environments where network reliability and security are paramount.

Open source communities often respond rapidly to security vulnerabilities, issuing patches and updates collaboratively.

Enabling Interoperability and Standards Alignment

Open source fosters adherence to open standards and protocols, ensuring interoperability among heterogeneous devices and systems. This is crucial in multi-vendor SDN deployments where seamless communication between controllers, switches, and AI modules is necessary.

Moreover, open source implementations often serve as reference platforms for standardization efforts, helping to shape future protocols and interfaces.

Community-Driven AI Solutions

The integration of AI into SDN management has been significantly propelled by open source initiatives, which provide accessible frameworks, tools, and collaborative environments for developing intelligent network solutions.

AI Frameworks and Toolkits

Foundational AI frameworks such as TensorFlow, PyTorch, and scikit-learn have democratized machine learning development. These open source platforms offer extensive libraries, pretrained models, and community support, enabling network engineers and researchers to build AI models tailored for SDN analytics, prediction, and automation without starting from scratch.

Open Source SDN Controllers with AI Capabilities

Several open source SDN controllers have embraced AI integration either natively or through extensible architectures:

- **OpenDaylight:** A modular and extensible SDN controller platform that supports integration with AI and machine learning tools. OpenDaylight's telemetry and analytics modules facilitate the collection and processing of network data for AI consumption.

- **ONOS (Open Network Operating System):** Designed for scalability and high availability, ONOS enables AI-driven network optimization through modular applications and APIs that allow AI models to influence routing, load balancing, and fault management.

- **Ryu:** A lightweight, Python-based SDN controller widely used in academic research and prototyping. Its simplicity and extensibility make it ideal for experimenting with AI-driven control algorithms.

Collaborative Datasets and Benchmarking Initiatives

Open source communities curate and share datasets that capture network traffic patterns, topology changes, and performance metrics. These datasets enable reproducible research, benchmarking of AI models, and fair evaluation of different approaches.

Examples include the CAIDA datasets and various SDN traffic traces published for research purposes.

AI-Orchestrated Network Automation Projects

Projects such as SONATA and ONAP exemplify how open source AI orchestrators automate complex network functions and lifecycle management. These platforms use AI to optimize resource allocation, service chaining, and fault detection, demonstrating real-world applicability of community-driven AI solutions in SDN environments.

Community Engagement and Knowledge Sharing

Open source communities foster vibrant ecosystems through mailing lists, forums, hackathons, and conferences. These venues encourage knowledge exchange, mentorship, and collaborative problem-solving, accelerating the maturation of AI-driven SDN technologies.

Table 22-1. *Notable Open Source Projects Supporting AI-Driven SDN Management*

Project Name	Description	AI Integration	Website
OpenDaylight	Modular SDN controller platform	AI telemetry and analytics plug-ins	`https://www.opendaylight.org`
ONOS	Scalable SDN OS for service providers	AI-enabled network optimization apps	`https://onosproject.org`
Ryu	Lightweight SDN controller framework	Extensible for AI research	`https://osrg.github.io/ryu`
TensorFlow	End-to-end open source ML framework	Model development and deployment	`https://www.tensorflow.org`
PyTorch	Deep learning framework with dynamic computation graphs	Flexible AI model prototyping	`https://pytorch.org`
SONATA	Network function virtualization platform	AI orchestration for automation	`https://www.sonata-nfv.eu/`
ONAP	Open network automation platform	AI-driven lifecycle management	`https://www.onap.org`

Future of Open Source in Networking

The trajectory of open source in AI-driven SDN management is promising, with emerging trends and technological advances poised to accelerate innovation and adoption.

Federated and Collaborative AI

Federated learning is gaining traction as a privacy-preserving approach where multiple organizations collaboratively train AI models on distributed data without sharing raw data. Open source frameworks supporting federated learning will enable robust, generalized AI models for SDN that respect data privacy and regulatory requirements.

This approach will foster cross-industry collaboration, accelerating AI model improvements and deployment.

Edge AI and Distributed Intelligence

As edge computing proliferates, open source projects will increasingly focus on deploying AI models at the network edge. Edge AI reduces latency and bandwidth consumption by processing data locally, enabling real-time SDN control and analytics.

Open source toolkits optimized for edge devices will empower decentralized intelligent network management, enhancing responsiveness and scalability.

Enhanced Automation and Intent-Based Networking

Intent-based networking abstracts network management by allowing operators to specify high-level business goals, which AI-driven systems automatically translate into network configurations.

Open source frameworks implementing intent-based networking will simplify management complexity, reduce human error, and enable adaptive networks that self-optimize based on changing conditions.

Standardization and Interoperability

Ongoing efforts to standardize AI interfaces, data models, and telemetry protocols will be complemented by open source reference implementations. This synergy ensures interoperability among diverse SDN components and AI modules, fostering a healthy ecosystem of interoperable tools and applications.

Growing Ecosystem and Community Engagement

The expansion of open source communities, educational initiatives, and collaborative research will continue to drive innovation and democratization. Hackathons, workshops, and online platforms will provide avenues for newcomers and experts alike to contribute and learn.

This vibrant ecosystem will accelerate the evolution of AI-driven SDN management, making it accessible to a broader audience and adaptable to diverse environments.

Conclusion

Open source initiatives are fundamental to democratizing AI-driven SDN management. By fostering innovation, reducing costs, and promoting transparency and interoperability, open source empowers a broad spectrum of stakeholders to participate in the evolution of intelligent networking.

Community-driven AI solutions unlock new possibilities for scalable, adaptive, and automated network control, enabling networks to meet the demands of increasingly complex digital infrastructures.

Looking forward, emerging trends such as federated learning, edge AI, and intent-based networking will be propelled by open source collaborations, shaping a future where AI-driven SDN management is not only powerful but also accessible, inclusive, and resilient.

CHAPTER 23

AI Agents for SD-WAN Optimization

Software-Defined Wide Area Networking (SD-WAN) has rapidly become a cornerstone technology for enterprises seeking flexible, cost-effective, and high-performance connectivity across geographically dispersed locations. By abstracting network control from hardware and enabling centralized, software-driven management, SD-WAN allows organizations to leverage multiple transport links—such as MPLS, broadband Internet, and cellular networks—in a unified and intelligent manner.

However, as the scale and complexity of SD-WAN deployments increase, traditional static configurations and manual management approaches struggle to keep pace with dynamic network conditions and evolving business requirements. Artificial intelligence (AI) agents offer a transformative solution by introducing automation, adaptability, and predictive capabilities that optimize SD-WAN performance continuously.

In the era of digital transformation, optimizing wide area networks is no longer a luxury but a necessity. AI agents empower SD-WAN to evolve beyond static management into an intelligent, adaptive system that antici-pates and responds to network demands in real time.

This chapter delves into the foundational concepts of SD-WAN, explores the AI techniques applied to its optimization, and outlines key performance metrics and evaluation frameworks. Readers will gain a comprehensive understanding of how AI agents enhance SD-WAN operations to deliver superior application performance, improved reliability, and operational efficiency.

The increasing reliance on cloud applications, remote workforces, and real-time services has put unprecedented demands on enterprise WANs. SD-WAN, with its software-centric approach, addresses many of these challenges by enabling dynamic

217

© Het Mehta 2026
H. Mehta, *AI Agents for Secure and Software-Defined Networking*,
https://doi.org/10.1007/979-8-8688-2358-9_23

path selection and centralized policy enforcement. Still, the sheer volume of data generated by SD-WAN devices, combined with the complexity of network policies and fluctuating link qualities, makes manual optimization impractical.

AI agents, equipped with advanced analytics and learning capabilities, can continuously monitor network telemetry, learn traffic patterns, predict potential issues, and autonomously adjust network configurations. This proactive approach not only improves network performance but also reduces operational costs and enhances security posture.

Introduction to SD-WAN

SD-WAN represents a paradigm shift in wide area networking, moving away from rigid, hardware-centric models toward flexible, software-defined architectures. This section introduces the core components, benefits, and challenges of SD-WAN to establish context for AI-driven optimization.

SD-WAN Architecture and Components

At its core, SD-WAN decouples the control plane from the data plane, enabling centralized orchestration and dynamic traffic management across diverse WAN links. The principal components include

- **Edge Devices:** Physical or virtual appliances deployed at branch offices, data centers, or cloud edges that handle traffic forwarding, policy enforcement, encryption, and telemetry collection. These devices are responsible for implementing the defined policies and ensuring secure and efficient data transmission.

- **Centralized Controller:** The management plane that defines policies, monitors network health, and orchestrates configurations across the SD-WAN fabric. It maintains a global view of the network topology and link statuses, enabling intelligent decision-making.

- **Orchestrator:** Provides administrators with an interface for provisioning, monitoring, and managing the network, automating deployment and updates. The orchestrator simplifies the complexity of managing multiple sites and diverse transport links.

- **Transport Links:** Multiple WAN connections such as MPLS, broadband Internet, LTE/5G, and satellite networks that carry data traffic. SD-WAN intelligently aggregates these links to optimize performance and cost.

- **Integrated Security Services:** Encryption, firewalling, segmentation, intrusion detection/prevention, and other security functions embedded within the SD-WAN solution to protect data integrity and privacy.

The architecture enables centralized policy control while allowing distributed packet forwarding, which is essential for scalability and performance.

Benefits of SD-WAN

The adoption of SD-WAN offers several compelling advantages:

- **Cost Optimization:** By leveraging inexpensive broadband and cellular links alongside or instead of costly MPLS circuits, organizations reduce WAN expenses significantly. This mix-and-match approach allows businesses to balance cost and performance effectively.

- **Application Performance:** Intelligent path selection ensures critical applications receive priority routing over optimal links, enhancing user experience. SD-WAN can identify application types and steer traffic accordingly, ensuring service level agreements (SLAs) are met.

- **Simplified Management:** Centralized control and automation reduce operational complexity and accelerate deployment of new sites or services. Administrators can define policies once and deploy them network-wide.

- **Agility and Scalability:** Rapid provisioning enables organizations to adapt quickly to changing business needs and scale their networks efficiently. This agility is vital in today's fast-paced digital economy.

- **Security Enhancements:** Embedded security features protect data in transit and enforce consistent policies across all locations, reducing the risk of breaches.

- **Improved Reliability:** SD-WAN can dynamically reroute traffic in response to link failures or degradations, maintaining continuous service availability.

Challenges in SD-WAN Management

Despite its benefits, SD-WAN presents operational challenges:

- **Heterogeneous Link Performance:** Diverse WAN links exhibit varying latency, jitter, bandwidth, and reliability, complicating traffic engineering. The variability requires continuous assessment to maintain optimal routing.

- **Dynamic Network Conditions:** Fluctuations in link quality due to congestion, outages, or interference require continuous monitoring and adaptation. Static policies may fail to respond adequately to such changes.

- **Complex Policy Management:** Managing consistent policies across distributed sites with diverse application requirements is challenging. Policy conflicts and misconfigurations can degrade performance or expose security risks.

- **Security Risks:** Expanding the attack surface with Internet-based links demands robust security controls and real-time threat detection.

- **Integration Complexity:** Coordinating SD-WAN with legacy infrastructure and multicloud environments requires sophisticated orchestration and interoperability.

- **Data Overload:** The massive volume of telemetry and logs generated can overwhelm human operators, making it difficult to extract actionable insights promptly.

These challenges underscore the importance of intelligent automation and adaptive control mechanisms—roles ideally fulfilled by AI agents.

AI Techniques for Optimization

AI agents employ a suite of advanced techniques to monitor, analyze, and optimize SD-WAN environments autonomously. This section discusses the key AI methodologies driving these capabilities.

Machine Learning for Traffic Classification and Prediction

Machine learning (ML) algorithms analyze network data to classify traffic flows, detect anomalies, and forecast future network states. These capabilities enable proactive and precise network management.

- **Traffic Classification:** Supervised learning models such as decision trees, support vector machines (SVM), random forests, and deep neural networks identify application types, user behaviors, and protocol patterns. Accurate classification enables tailored routing and quality of service (QoS) policies, ensuring critical applications receive priority.

- **Traffic Prediction:** Time series forecasting models—including Autoregressive Integrated Moving Average (ARIMA), Long Short-Term Memory (LSTM) networks, and Facebook's Prophet—predict bandwidth demand and traffic patterns. Anticipating traffic surges allows the network to allocate resources proactively, avoiding congestion.

- **Anomaly Detection:** Unsupervised algorithms like clustering, autoencoders, and isolation forests detect deviations from normal traffic behavior, signaling faults, security threats, or misconfigurations. Early detection facilitates rapid remediation.

- **Feature Engineering and Data Quality:** The effectiveness of ML models depends heavily on selecting relevant features from raw telemetry and ensuring high-quality, labeled datasets for training.

- **Continuous Learning:** Models must adapt over time to evolving traffic patterns and network conditions, requiring ongoing retraining and validation.

Reinforcement Learning for Dynamic Path Selection and Load Balancing

Reinforcement learning (RL) enables AI agents to learn optimal routing policies by interacting with the network environment and receiving feedback based on performance outcomes. Training RL agents requires accurate network simulation environments. Digital twins (Chapter 5) or network emulators (Mininet, NS3, GNS3) provide realistic, safe testing grounds where agents can learn optimal provisioning strategies.

- **Adaptive Routing:** RL agents dynamically select paths to minimize latency, packet loss, and congestion, adapting to real-time network conditions. By trial and error, the agent discovers routing strategies that maximize network performance.

- **Multi-agent Systems:** Collaborative RL agents manage different network segments or layers, enabling scalable and distributed control. Agents can coordinate to optimize global network objectives.

- **Reward Functions:** Designing appropriate reward signals is critical to guide learning toward desired outcomes, balancing throughput, delay, cost, and reliability.

- **Challenges:** Balancing exploration of new paths with exploitation of known good routes requires careful tuning; training can be resource-intensive and may involve risks during learning phases.

- **Simulation-Based Training:** Due to the risks of live network experimentation, RL agents are often trained in simulated environments before deployment.

Optimization Algorithms for Resource Allocation

Optimization techniques solve mathematical problems to allocate bandwidth and prioritize traffic efficiently, complementing ML and RL approaches.

- **Genetic Algorithms:** Inspired by natural selection, these evolutionary algorithms iteratively improve candidate solutions for resource allocation and traffic prioritization.

- **Simulated Annealing:** A probabilistic method that explores the solution space to avoid local optima, suitable for complex, nonlinear optimization problems.

- **Gradient-Based Methods:** Used in differentiable models to fine-tune policy parameters through gradient descent and related techniques.

- **Linear and Nonlinear Programming:** Formulate resource allocation as constrained optimization problems solvable by established mathematical solvers.

- **Multi-objective Optimization:** Balances competing goals such as minimizing latency, maximizing throughput, and reducing operational costs.

- **Real-Time Constraints:** Optimization algorithms must operate within strict time limits to be effective in dynamic SD-WAN environments.

Natural Language Processing for Intent-Based Policy Management

Natural Language Processing (NLP) allows AI agents to interpret human operator intents expressed in natural language and translate them into precise network policies.

- **Intent Extraction:** Semantic parsing, named entity recognition, and dependency parsing identify key commands, parameters, and constraints from operator input.

- **Policy Generation:** The AI converts intents into routing, security, and QoS configurations compatible with the SD-WAN controller.

- **Dialogue Systems:** Interactive interfaces enable operators to refine or clarify intents through conversational exchanges.

- **Benefits:** Simplifies network management, reduces configuration errors, and accelerates policy deployment by enabling nonexpert users to define network behavior intuitively.

- **Limitations:** Ambiguity, context understanding, and domain-specific terminology pose challenges; continuous improvement and domain adaptation are required.

- **Integration with Automation:** NLP interfaces are often combined with automated validation and simulation to ensure safe deployment.

AI-Driven Fault Detection and Self-Healing

AI agents analyze telemetry data to detect faults early, diagnose root causes, and initiate automated remediation actions to maintain network health.

- **Predictive Maintenance:** Machine learning models identify degradation trends and predict failures before they impact service.

- **Root Cause Analysis:** Correlation analysis, causal inference, and graph-based techniques pinpoint problematic devices, links, or configurations.

- **Automated Recovery:** AI triggers corrective actions such as traffic rerouting, device reboot, or configuration rollback without human intervention.

- **Alert Prioritization:** AI filters and prioritizes alerts to reduce operator overload and focus attention on critical issues.

- **Continuous Feedback:** Self-healing actions are monitored and evaluated to improve future responses.

- **Resilience Improvement:** These capabilities reduce mean time to repair (MTTR) and enhance overall network availability.

Performance Metrics and Evaluation

Measuring the effectiveness of AI-driven SD-WAN optimization requires comprehensive metrics and evaluation strategies to validate improvements and guide ongoing enhancements.

Key Performance Metrics

- **Latency:** Measures the round-trip time for packets between endpoints. Low latency is essential for real-time applications such as voice over IP (VoIP) and video conferencing.

- **Packet Loss:** The percentage of packets lost during transmission. High packet loss degrades application quality and user experience.

- **Throughput:** The amount of data successfully transmitted over the network per unit time. Higher throughput indicates better network capacity utilization.

- **Jitter:** Variation in packet delay, which can disrupt streaming and VoIP services, causing choppy audio or video.

- **Link Utilization:** Reflects how efficiently bandwidth is used across available transport links. Balanced utilization avoids congestion and underutilization.

- **Cost Efficiency:** Quantifies reductions in network operating expenses achieved by optimizing link usage and traffic routing.

- **Application Performance:** End-user experience metrics such as page load times, transaction completion rates, and video resolution.

- **Adaptability and Responsiveness:** Measures how quickly and effectively AI agents respond to network changes, faults, or security incidents.

- **Reliability and Availability:** Uptime and fault tolerance of the SD-WAN infrastructure, reflecting service continuity.

Evaluation Methodologies

- **Simulation and Emulation:** Tools like ns-3, Mininet, and GNS3 simulate network topologies and traffic to test AI algorithms under controlled conditions. These environments allow safe experimentation without impacting production networks.

- **Testbed Deployments:** Physical or virtual testbeds replicate real-world environments for validating AI-driven optimization in near-production settings.

- **Benchmarking Against Baselines:** Comparing AI-enhanced SD-WAN performance with traditional static routing or heuristic approaches to quantify improvements.

- **A/B Testing:** Running parallel network configurations to measure user experience improvements directly and objectively.

- **Longitudinal Monitoring:** Continuous telemetry collection post-deployment assesses sustained performance and informs model retraining.

- **User Feedback:** Incorporating subjective user experience data complements objective metrics for holistic evaluation.

Visualization and Reporting

Effective visualization of network performance and AI agent decisions is crucial for operator trust and collaboration.

- **Dashboards:** Present real-time metrics, alerts, and trend analyses in intuitive formats.

- **Explainability Tools:** Provide insights into AI decision-making processes, helping operators understand and validate actions.

- **Automated Reporting:** Generates periodic performance summaries to support compliance, auditing, and strategic planning.

- **Interactive Analytics:** Enables drill-down into specific events or metrics for root cause investigation.

Table 23-1. *Summary of AI Techniques Applied to SD-WAN Optimization*

AI Technique	Application Area	Benefits	Challenges
Machine Learning	Traffic classification, prediction	Proactive resource allocation, anomaly detection	Requires quality data, model training overhead
Reinforcement Learning	Dynamic path selection, load balancing	Adaptive routing, self-learning policies	Exploration–exploitation trade-off, training complexity
Optimization Algorithms	Bandwidth allocation, policy tuning	Efficient resource utilization, multi-objective balancing	Computational complexity, convergence issues
Natural Language Processing	Policy interpretation, intent translation	Simplifies configuration, operator-friendly	Ambiguity in language, contextual understanding
AI-Driven Fault Detection	Self-healing and resilience	Early fault detection, automated recovery	False positives, extensive telemetry requirements

Case Study: AI Agents in Action for SD-WAN Optimization

Consider a multinational enterprise with over 200 branch offices connected via a mix of MPLS and broadband links. The organization faced challenges with fluctuating link quality, inconsistent application performance, and rising WAN costs.

By deploying AI agents integrated with their SD-WAN controller, the enterprise achieved

- **Dynamic Traffic Steering:** Reinforcement learning agents continuously optimized path selection, reducing average latency by 25%, improving voice and video call quality.

- **Predictive Bandwidth Allocation:** Machine learning models forecasted traffic spikes, enabling preemptive resource adjustments that minimized packet loss during peak hours.

- **Automated Fault Recovery:** AI-driven anomaly detection identified link degradations early, triggering automated rerouting and reducing downtime by 40%.

- **Simplified Policy Management:** NLP interfaces allowed network administrators to express high-level intents, accelerating policy updates and reducing configuration errors.

- **Cost Savings:** Optimized use of broadband links reduced MPLS dependency, lowering WAN expenses by 30%.

This case demonstrates how AI agents transform SD-WAN operations, delivering measurable improvements in performance, reliability, and operational efficiency.

Future Directions in AI-Driven SD-WAN Optimization

The integration of AI in SD-WAN is evolving rapidly, with several promising trends on the horizon:

- **Federated Learning:** Collaborative AI model training across multiple organizations without sharing sensitive data will enhance model robustness and privacy, enabling better generalization across diverse networks.

- **Edge AI:** Deploying AI inference capabilities on edge devices will enable ultra-low latency decision-making and reduce dependence on centralized controllers, improving responsiveness and scalability.

- **Explainable AI (XAI):** Enhancing transparency of AI decisions will build operator trust and facilitate regulatory compliance, addressing concerns about AI "black boxes."

- **Integration with 5G and IoT:** AI agents will optimize SD-WAN in complex environments involving 5G connectivity and massive IoT deployments, managing diverse traffic profiles and stringent latency requirements.

- **Cross-Domain Orchestration:** AI will coordinate SD-WAN with other network domains such as data center and access networks for holistic optimization, enabling end-to-end service assurance.

- **Security Enhancements:** AI-driven threat detection and response will become integral to SD-WAN, protecting increasingly complex and distributed networks.

Conclusion

AI agents are revolutionizing SD-WAN optimization by introducing intelligence, automation, and adaptability into network management. Through machine learning, reinforcement learning, optimization algorithms, and natural language processing, AI agents enable dynamic traffic steering, efficient resource allocation, proactive fault detection, and simplified policy management.

Evaluating these AI-driven enhancements through comprehensive metrics and rigorous methodologies ensures that SD-WAN deployments achieve improved performance, reliability, and cost-effectiveness. Real-world deployments validate the transformative impact of AI integration.

As SD-WAN continues to evolve amid growing demands for cloud connectivity, remote work, and real-time services, AI agents will be indispensable in managing complexity and delivering seamless network experiences across distributed enterprises.

CHAPTER 24

AI-Driven Security for SD-WAN

The rapid adoption of Software-Defined Wide Area Networking (SD-WAN) has revolutionized enterprise connectivity by offering flexibility, cost efficiency, and improved performance. However, this architectural shift introduces new security challenges that traditional protection mechanisms struggle to address. SD-WAN environments, characterized by distributed edge devices, multiple transport links, and cloud integration, expand the attack surface and increase exposure to sophisticated cyber threats.

Artificial intelligence (AI) emerges as a critical enabler for securing SD-WAN, providing intelligent, adaptive, and automated defense capabilities. By leveraging AI's power to analyze vast amounts of network data, detect anomalies, and predict potential attacks, organizations can enhance their security posture while maintaining agility and performance.

This chapter explores the security challenges inherent in SD-WAN architectures, presents AI-based solutions that strengthen defense mechanisms, and examines real-world case studies demonstrating the effectiveness of AI-driven security in SD-WAN deployments. Readers will gain insights into how AI agents can safeguard modern networks against evolving threats while enabling secure digital transformation.

Security Challenges in SD-WAN

While SD-WAN offers numerous operational benefits, it also introduces unique security concerns that must be addressed to protect enterprise data and infrastructure.

H. Mehta, *AI Agents for Secure and Software-Defined Networking*,
https://doi.org/10.1007/979-8-8688-2358-9_24

Expanded Attack Surface

SD-WAN's use of multiple transport links, including public Internet and cellular networks, increases exposure to external threats. Unlike traditional MPLS-based WANs with closed circuits, SD-WAN traffic traverses less secure paths, making interception, man-in-the-middle attacks, and unauthorized access more likely.

Distributed Edge Vulnerabilities

Edge devices deployed at branch offices or cloud edges often have less physical security and may be managed remotely, increasing the risk of device tampering, misconfiguration, or compromise. Attackers can exploit these vulnerabilities to gain footholds within the network.

Complex Policy Management

The dynamic and distributed nature of SD-WAN requires consistent enforcement of security policies across diverse environments. Managing and auditing these policies manually is error-prone, leading to potential gaps or conflicts that attackers can exploit.

Encryption and Key Management Challenges

SD-WAN relies heavily on encryption to secure data in transit. However, managing certificates, keys, and cryptographic protocols at scale across distributed devices presents operational complexity and risk of mismanagement.

Threat Detection in Encrypted Traffic

While encryption protects data confidentiality, it also obscures traffic content from conventional security inspection tools, complicating threat detection and intrusion prevention efforts. AI analyzes metadata and statistical features—such as packet sizes, inter-arrival times, flow duration, and connection patterns—to infer application types and detect threats without the computational burden of full decryption, balancing security visibility with privacy preservation and performance.

Insider Threats and Lateral Movement

Compromised internal users or devices can leverage SD-WAN's interconnected fabric to move laterally across the network, escalating privileges and exfiltrating sensitive data.

Integration with Cloud and Third-Party Services

SD-WAN often integrates with multiple cloud providers and third-party applications, introducing supply chain risks and complicating security monitoring and policy enforcement.

AI Solutions for Enhanced Security

AI-driven security solutions address the complex challenges of SD-WAN by delivering proactive, adaptive, and automated defenses that evolve with the threat landscape.

Anomaly Detection and Behavioral Analytics

AI agents utilize machine learning algorithms to establish baseline behaviors for users, devices, and network traffic. Deviations from these baselines—such as unusual access patterns, data transfers, or communication flows—trigger alerts or automated responses.

- **Unsupervised Learning:** Techniques like clustering and autoencoders detect unknown threats without labeled data.

- **Behavioral Profiling:** AI models continuously refine user and device profiles to improve detection accuracy.

- **Example:** Detecting a sudden surge in data uploads from a branch office outside normal business hours may indicate data exfiltration.

Threat Intelligence Integration

AI systems ingest global threat intelligence feeds and correlate them with local network data to identify known malicious IPs, domains, or attack signatures.

- **Real-Time Updates:** Continuous ingestion ensures defenses stay current against emerging threats.

- **Contextual Analysis:** AI evaluates the relevance of threat intelligence based on network context, reducing false positives.

Automated Incident Response and Mitigation

Upon detecting threats, AI agents can initiate immediate countermeasures without human intervention, minimizing damage and response times.

- **Traffic Quarantine:** Isolating suspicious devices or flows

- **Dynamic Policy Adjustment:** Modifying firewall rules or access controls in real time

- **Forensic Data Collection:** Capturing evidence for post-incident analysis

- **Self-Healing:** Restarting compromised services or rerouting traffic away from affected links

Encryption and Key Management Automation

AI assists in managing cryptographic assets by automating certificate issuance, renewal, revocation, and key rotation.

- **Risk-Based Prioritization:** AI evaluates device criticality and threat exposure to prioritize key management tasks.

- **Anomaly Detection:** Identifies unusual certificate usage or unauthorized key access.

- **Reduced Human Error:** Automation minimizes misconfigurations that could lead to vulnerabilities.

Deep Packet Inspection and Encrypted Traffic Analysis

Advanced AI techniques analyze metadata, traffic patterns, and statistical features to detect threats within encrypted flows without decrypting content.

- **Traffic Fingerprinting:** Identifies application types and anomalous behaviors

- **Side-Channel Analysis:** Uses timing, packet size, and flow characteristics to infer malicious activity

- **Privacy Preservation:** Enables security monitoring while maintaining data confidentiality

Insider Threat Detection and Access Control

AI models monitor user behavior and access patterns to detect potential insider threats and enforce least privilege principles.

- **User and Entity Behavior Analytics (UEBA):** Identifies deviations such as privilege escalations or unusual resource access

- **Adaptive Access Policies:** Dynamically adjust permissions based on risk assessments

Integration with Security Orchestration, Automation, and Response (SOAR)

AI-powered SD-WAN security solutions integrate with broader SOAR platforms to coordinate multi-layered defenses across the enterprise.

- **Automated Playbooks:** Define standardized response workflows triggered by AI detections.

- **Cross-Domain Correlation:** Combines SD-WAN data with endpoint, cloud, and SIEM information for comprehensive threat management.

Case Studies in SD-WAN Security

The following real-world examples illustrate how AI-driven security enhances SD-WAN deployments, delivering measurable benefits in threat detection, response, and overall network resilience.

Financial Services Firm Secures Branch Networks

A multinational financial institution deployed an AI-enhanced SD-WAN solution to protect its 150+ branch offices. Key outcomes included

- **Early Threat Detection:** Behavioral analytics identified insider threats involving unauthorized data access, enabling timely intervention.

- **Automated Policy Enforcement:** AI dynamically adjusted firewall rules based on detected threats, reducing manual configuration errors.

- **Compliance Reporting:** Automated evidence collection simplified audits for regulatory compliance.

Healthcare Provider Protects Patient Data

A large healthcare provider leveraged AI-driven security to safeguard sensitive patient information transmitted over SD-WAN links connecting hospitals and clinics.

- **Encrypted Traffic Analysis:** AI detected malware communications hidden within encrypted traffic without violating privacy regulations.

- **Incident Response Automation:** Rapid isolation of compromised devices minimized outbreak impact.

- **Key Management Automation:** Ensured continuous encryption without service disruption.

Retail Chain Enhances Security at Scale

A global retail chain with thousands of outlets employed AI agents to secure its SD-WAN infrastructure supporting point-of-sale systems and inventory management.

- **Threat Intelligence Correlation:** AI filtered millions of threat signals daily, focusing on relevant risks to retail operations.

- **Self-Healing Capabilities:** Automated remediation reduced mean time to recovery (MTTR) by 50%.

- **Scalable Security Management:** Centralized AI-driven orchestration simplified policy updates across diverse locations.

Table 24-1. *AI-Driven Security Techniques and Their Benefits in SD-WAN*

Technique	Description	Benefits	Challenges
Behavioral Analytics	Detects anomalies based on learned patterns	Early threat detection, reduced false positives	Requires quality baseline data
Threat Intelligence Integration	Correlates global threats with local data	Up-to-date defenses, contextual awareness	Data overload, relevance filtering
Automated Incident Response	AI-triggered mitigation actions	Rapid containment, reduced downtime	Risk of false positives triggering actions
Encryption and Key Management	Automates cryptographic asset lifecycle	Enhanced security, reduced errors	Complexity in heterogeneous environments
Encrypted Traffic Analysis	Analyzes metadata without decryption	Privacy-preserving threat detection	Limited visibility compared to full inspection
Insider Threat Detection	Monitors user behavior and access	Mitigates internal risks	Privacy concerns, behavioral variability
SOAR Integration	Coordinates multi-layered security responses	Comprehensive threat management	Integration complexity

Conclusion

Securing SD-WAN environments is a complex and evolving challenge, driven by the architecture's distributed nature, diverse transport links, and integration with cloud services. AI-driven security solutions provide a powerful arsenal to address these challenges by delivering intelligent threat detection, real-time response, and automated policy management.

Through behavioral analytics, threat intelligence integration, incident automation, and advanced encrypted traffic analysis, AI agents enhance the resilience and trustworthiness of SD-WAN deployments. Real-world case studies demonstrate that organizations leveraging AI in SD-WAN security achieve faster detection, reduced operational overhead, and improved compliance.

As cyber threats continue to grow in sophistication, the synergy between AI and SD-WAN will be essential for building secure, agile, and future-ready networks. Embracing AI-driven security is not only a strategic advantage but a necessity for safeguarding enterprise digital transformation.

CHAPTER 25

Ensuring Business Continuity with AI-Driven Disaster Recovery

In the preceding chapters, we explored how AI fortifies SD-WAN environments by enhancing security through proactive threat detection and automated mitigation. While these capabilities significantly reduce the risk of disruptions, no system is completely immune to failures or disasters. Natural calamities, hardware malfunctions, cyberattacks, or human errors can still cause significant downtime.

This reality necessitates a robust disaster recovery (DR) strategy that ensures business continuity by enabling rapid restoration of critical systems and data. Traditional DR approaches, however, often fall short in today's fast-paced, distributed SD-WAN networks due to their reliance on manual processes and static procedures.

> *In the face of uncertainty, resilience is the hallmark of successful enterprises. AI-driven disaster recovery not only safeguards data but ensures that business operations continue seamlessly, turning potential crises into manageable challenges.*

Artificial intelligence offers a transformative approach to disaster recovery by automating complex workflows, predicting failures before they occur, and dynamically optimizing resource allocation during recovery. This chapter delves into the essentials of disaster recovery planning, AI-driven recovery techniques, SD-WAN-specific recovery strategies, and real-world best practices. Together, these insights provide a comprehensive framework for resilient, AI-enabled business continuity.

© Het Mehta 2026
H. Mehta, *AI Agents for Secure and Software-Defined Networking*,
https://doi.org/10.1007/979-8-8688-2358-9_25

Disaster Recovery Planning Essentials

Before integrating AI into disaster recovery, organizations must establish a solid foundation through careful planning. This section outlines the key components of effective DR planning.

Risk Assessment and Business Impact Analysis

The first step in disaster recovery planning is conducting a thorough risk assessment to identify potential threats such as cyberattacks, hardware failures, or environmental disasters. Alongside this, a business impact analysis (BIA) evaluates which systems and processes are critical, estimating the financial and operational consequences of downtime.

Key concepts:

- **Recovery Time Objective (RTO):** The maximum acceptable downtime for a system

- **Recovery Point Objective (RPO):** The maximum acceptable data loss measured in time

Understanding these metrics guides prioritization in recovery efforts.

Defining Recovery Strategies

Once risks and impacts are understood, organizations select appropriate recovery strategies. These may include

- **On-Premises Backups:** Local copies of data and configurations

- **Secondary Data Centers:** Physical or virtual sites for failover

- **Cloud-Based Disaster Recovery as a Service (DRaaS):** Leveraging cloud scalability for rapid recovery

Each approach involves trade-offs in cost, complexity, and recovery speed.

Documentation and Procedures

Comprehensive documentation is vital. Recovery playbooks should detail step-by-step procedures, roles and responsibilities, and escalation paths. Clear communication protocols must be established for internal teams, customers, and partners to ensure coordinated response.

Testing and Continuous Improvement

Regular testing through simulations and drills validates the DR plan's effectiveness. Testing uncovers weaknesses and areas for improvement, which should be incorporated into plan revisions. Continuous improvement ensures preparedness evolves with changing technologies and threats.

Transition to AI Techniques: While traditional DR planning forms the backbone of recovery readiness, the complexity and speed required in modern distributed SD-WAN environments demand more intelligent, automated solutions. The following section explores how AI techniques revolutionize disaster recovery by enhancing prediction, automation, and adaptability.

AI Techniques for Recovery

Artificial intelligence brings advanced capabilities to disaster recovery, transforming it from a reactive process into a proactive, adaptive system.

Predictive Analytics for Risk Mitigation

AI-powered predictive analytics analyze historical incident data, system logs, and real-time telemetry to forecast potential failures or cyber threats. Early warnings enable preventive actions that can avoid or mitigate disasters altogether.

Automated Recovery Orchestration

AI-driven orchestration platforms automate the execution of complex recovery workflows, coordinating multiple steps such as failover, data restoration, and network reconfiguration without human intervention. This automation reduces recovery time and human error.

Dynamic Resource Allocation

During recovery, AI dynamically allocates computing, storage, and network resources to meet defined RTO and RPO targets efficiently. This optimization prevents bottlenecks and ensures that resources are used cost-effectively.

Anomaly Detection During Recovery

Continuous AI monitoring detects anomalies or failures during recovery processes, triggering alerts or automated corrective actions to maintain recovery integrity and security compliance.

Intelligent Testing and Validation

AI simulates diverse disaster scenarios and dynamically adjusts test parameters based on system responses. This intelligent testing improves plan validation, ensuring recovery procedures remain effective as infrastructure evolves.

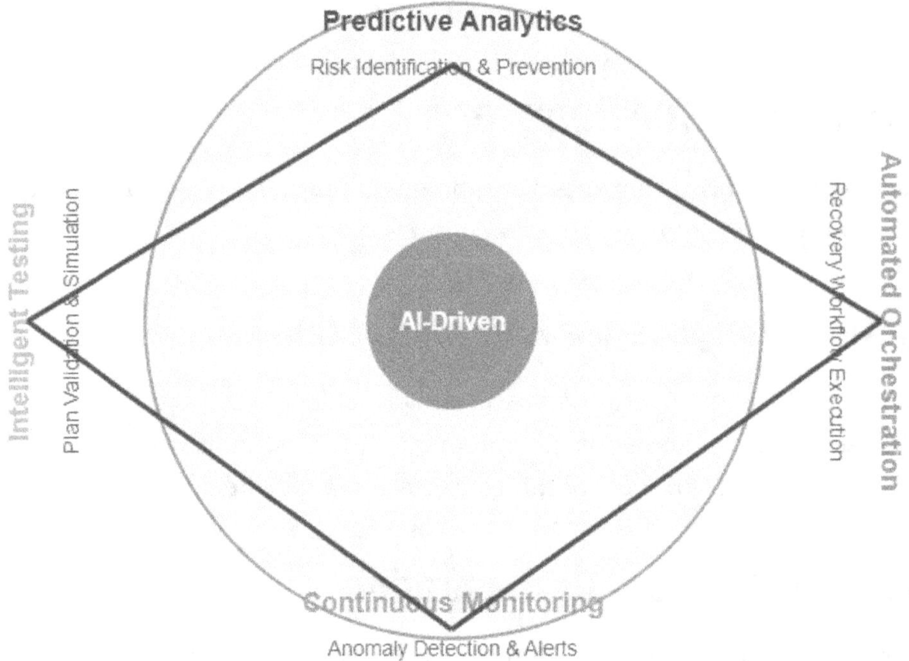

Figure 25-1. *AI-Driven Disaster Recovery Process in SD-WAN Environments*

SD-WAN-Specific Recovery Approaches

Building on the AI-driven recovery techniques, this section focuses on how these capabilities apply uniquely to SD-WAN environments, which present their own challenges and opportunities.

Multi-path Failover and Traffic Rerouting

SD-WAN supports multiple transport paths including MPLS, broadband, and cellular connections. AI continuously monitors link quality and performance, enabling rapid rerouting of traffic to avoid outages or degraded links. This multi-path failover capability is essential for maintaining connectivity during disasters.

Edge Device State Preservation

SD-WAN edge devices hold critical configurations and session states that must be preserved for consistent network behavior. AI automates backup and restoration of these distributed devices, reducing recovery time and preventing configuration drift.

Cloud-Integrated Recovery

With SD-WAN's inherent cloud connectivity, AI orchestrates failover to cloud-hosted applications and data stores when on-premises infrastructure is compromised, ensuring business services remain uninterrupted.

Security-Aware Recovery

AI ensures that recovery actions comply with security policies and monitors for threats that could exploit vulnerabilities during failover or restoration, maintaining a secure environment throughout the recovery process.

Transition to Real-World Examples: The practical benefits of AI-driven disaster recovery in SD-WAN are best understood through real-world applications. The following examples illustrate successful implementations and lessons learned.

Real-World Examples and Best Practices

Global Manufacturing Company

A multinational manufacturer integrated AI-driven disaster recovery with its SD-WAN infrastructure to protect critical supply chain systems. Predictive analytics detected network anomalies indicating impending hardware failures, triggering automated failover to backup sites. This reduced production downtime by 60% and optimized resource utilization during recovery.

Financial Services Provider

A financial institution deployed AI orchestration to automate recovery workflows across multiple data centers and cloud environments. AI-managed resource allocation ensured compliance with strict RTO and RPO targets. Regular AI-driven testing uncovered configuration inconsistencies, improving recovery plan reliability and reducing recovery times.

Healthcare Network

A healthcare provider used AI to simulate disaster scenarios within its SD-WAN-connected hospitals and clinics. Intelligent testing improved the realism and coverage of recovery drills, while AI-monitored recovery processes detected and corrected deviations in real time, ensuring uninterrupted patient care.

Table 25-1. *Key Benefits of AI-Driven Disaster Recovery*

Benefit	Description	Business Impact
Predictive Failure Detection	Anticipates disruptions before they occur	Minimizes unplanned downtime
Automated Orchestration	Executes recovery workflows without manual intervention	Speeds recovery and reduces errors
Dynamic Resource Optimization	Efficiently allocates resources during recovery	Meets recovery objectives cost-effectively
Continuous Monitoring	Detects anomalies during recovery	Maintains process integrity
Intelligent Testing	Simulates realistic scenarios and validates plans	Enhances preparedness and plan quality

Conclusion

Ensuring business continuity in today's complex SD-WAN environments requires disaster recovery strategies that are intelligent, automated, and adaptive. AI-driven disaster recovery transforms traditional approaches by providing predictive insights, orchestrating recovery workflows autonomously, and optimizing resource utilization.

By integrating AI techniques tailored to SD-WAN's unique characteristics, organizations can reduce downtime, maintain security during recovery, and ensure rapid restoration of critical services. The real-world examples in this chapter demonstrate that AI-enabled disaster recovery is not only a technological advancement but a strategic imperative for resilient, future-ready enterprises.

Together with AI-driven security, AI-powered disaster recovery completes a comprehensive framework for safeguarding modern networks—enabling businesses to anticipate threats, respond swiftly, and sustain continuous operations regardless of disruptions.

CHAPTER 26

Additional Resources and Glossary

Empowering readers with curated knowledge and clear definitions ensures deeper understanding and continued growth beyond this book.

As the fields of Artificial Intelligence (AI) and Software-Defined Networking (SDN) rapidly evolve, continuous learning is essential to stay current with emerging technologies, best practices, and research breakthroughs. This chapter provides a curated collection of additional resources—books, blogs, online courses, research papers, and tools—to support further exploration. It also includes a glossary of 50 important terms frequently encountered in AI-driven networking, offering clear and concise definitions to aid comprehension.

Recommended Books

- *Software Defined Networking: Design and Deployment* by Patricia A. Morreale and James M. Anderson: A comprehensive guide on SDN architecture, protocols, and deployment strategies

- *Artificial Intelligence: A Modern Approach* by Stuart Russell and Peter Norvig: The definitive textbook covering AI fundamentals and advanced topics

- *Deep Learning* by Ian Goodfellow, Yoshua Bengio, and Aaron Courville: An authoritative resource on deep learning theory and applications

© Het Mehta 2026
H. Mehta, *AI Agents for Secure and Software-Defined Networking,*
https://doi.org/10.1007/979-8-8688-2358-9_26

- *Machine Learning for Networking: Techniques and Applications* by Fei Hu: Focuses on applying machine learning techniques to network management and optimization

- *Network Function Virtualization (NFV) with a Touch of SDN* by Rajendra Chayapathi, Syed Farrukh Hassan, and Paresh Shah: Explores NFV and SDN integration with real-world examples

- *Reinforcement Learning: An Introduction* by Richard S. Sutton and Andrew G. Barto: A foundational text on reinforcement learning principles and algorithms

Influential Blogs and Websites

- **SDN Central** (`https://www.sdncentral.com/`): Offers industry news, tutorials, and analysis on SDN and network virtualization

- **The AI Network** (`https://theainetwork.com/`): Provides insights into AI applications in networking and telecommunications

- **Google AI Blog** (`https://ai.googleblog.com/`): Features cutting-edge AI research and projects from Google, including networking applications

- **Cisco DevNet** (`https://developer.cisco.com/`): Developer resources, APIs, and learning labs for SDN and AI-powered networking

- **Open Networking Foundation (ONF)** (`https://opennetworking.org/`): Provides standards, whitepapers, and community projects on SDN and programmable networks

Online Courses and Tutorials

- **AI For Everyone** by Andrew Ng (Coursera): A nontechnical introduction to AI concepts and applications (`https://www.coursera.org/learn/ai-for-everyone`)

- **Software Defined Networking** by Georgia Tech (edX): Comprehensive SDN fundamentals and hands-on labs (`https://www.edx.org/course/software-defined-networking`)

- **Machine Learning Engineer Nanodegree** (Udacity): Deep dive into machine learning algorithms with projects (`https://www.udacity.com/course/machine-learning-engineer-nanodegree--nd009t`)

- **Reinforcement Learning Foundations** (LinkedIn Learning): Introductory course on reinforcement learning concepts and applications (`https://www.linkedin.com/learning/reinforcement-learning-foundations`)

- **Network Automation with Python and Ansible** (Pluralsight): Practical automation skills for network engineers (`https://www.pluralsight.com/courses/network-automation-python-ansible`)

Key Research Papers and Reports

- "A Survey on Software-Defined Networking," Kreutz et al., *IEEE Communications Surveys & Tutorials*, 2015

- "Deep Reinforcement Learning for Network Optimization," Mao et al., *IEEE Communications Magazine*, 2019

- "Federated Learning: Challenges, Methods, and Future Directions," Li et al., *IEEE Signal Processing Magazine*, 2020

- "AI-Driven Network Automation and Orchestration," *Open Networking Foundation Whitepaper*, 2021

- "Explainable AI: Interpreting, Explaining and Visualizing Deep Learning," Samek et al., 2017

These papers provide foundational and advanced insights into AI and SDN integration.

Useful Tools and Libraries

- **TensorFlow**: An open source platform for machine learning and deep learning (`https://www.tensorflow.org/`)

- **PyTorch**: A flexible deep learning framework with dynamic computation graphs (`https://pytorch.org/`)

- **Mininet**: A network emulator for SDN prototyping and testing (`http://mininet.org/`)

- **ONOS**: Open source SDN network operating system with AI integration capabilities (`https://onosproject.org/`)

- **OpenDaylight**: Modular open source SDN controller platform (`https://www.opendaylight.org/`)

- **Scikit-learn**: Python library for classical machine learning algorithms (`https://scikit-learn.org/`)

- **Reinforcement Learning Coach**: RL framework for training and evaluation (`https://nervanasystems.github.io/coach/`)

Glossary of Important Terms

- **AI Agent:** Autonomous software entity capable of sensing its environment, making decisions, and acting to achieve goals

- **Anomaly Detection:** Identifying patterns in data that deviate from expected behavior

- **Autonomous Network:** A network capable of self-management with minimal human intervention

- **Bandwidth:** Maximum data transfer rate of a network connection

- **Big Data:** Extremely large datasets analyzed computationally to reveal patterns and trends

- **Cloud Computing:** Delivery of computing services over the Internet on a pay-as-you-go basis

- **Controller:** Centralized component in SDN that manages network behavior and policies

- **Data Plane:** Network layer responsible for forwarding data packets

- **Deep Learning:** Subset of machine learning using neural networks with multiple layers

- **Digital Twin:** Virtual replica of a physical system used for simulation and analysis

- **Edge Computing:** Processing data near the source to reduce latency and bandwidth use

- **Federated Learning:** Collaborative machine learning without sharing raw data among participants

- **Flow Table:** Data structure in SDN switches that stores forwarding rules

- **Graph Neural Network (GNN):** Neural network designed to operate on graph-structured data

- **Intent-Based Networking (IBN):** Network management approach where high-level business intents are translated into network policies

- **Latency:** Time delay experienced in data communication

- **Machine Learning (ML):** Algorithms that improve automatically through experience and data

- **Multi-agent System:** System composed of multiple interacting intelligent agents

- **Network Function Virtualization (NFV):** Virtualizing network services traditionally run on hardware appliances

- **OpenFlow:** Protocol enabling SDN controllers to communicate with switches

- **Policy Management:** Defining and enforcing rules that govern network behavior

- **Programmable Data Plane:** Network devices whose packet processing behavior can be programmed dynamically

- **Quality of Service (QoS):** Mechanisms to prioritize network traffic to meet performance requirements

- **Reinforcement Learning (RL):** Learning optimal actions through trial and error with feedback

- **SDN Controller:** Software entity that centrally manages SDN network behavior

- **Service Function Chaining (SFC):** Ordered sequence of network services applied to traffic flows

- **Software-Defined Network (SDN):** Network architecture separating control and data planes for programmability

- **Telemetry:** Automated collection and transmission of network data for monitoring

- **Traffic Engineering:** Optimizing the flow of data across a network to improve performance

- **Virtual Network Function (VNF):** Software implementation of a network function running on virtualized infrastructure

- **Zero-Touch Provisioning (ZTP):** Automated device onboarding and configuration without manual intervention

- **Anomaly:** An event or observation that deviates from the norm

- **Behavioral Analytics:** Analysis of user or system behavior to detect unusual activities

- **Controller Scalability:** Ability of SDN controllers to handle increasing network sizes and loads

- **Deep Reinforcement Learning (DRL):** Combination of deep learning and reinforcement learning techniques

- **Explainable AI (XAI):** AI methods that provide human-understandable explanations for decisions

- **Flow Setup Latency:** Time taken to install forwarding rules for new flows

- **Intent Translation:** Converting high-level user/business intents into network policies

- **Machine Learning Model:** Mathematical model trained to perform specific tasks based on data

- **Multi-tenancy:** Multiple users or tenants sharing a common infrastructure with isolation

- **Network Orchestration:** Automated coordination of network resources and services

- **Predictive Analytics:** Using data, statistical algorithms, and ML to predict future events

- **Resource Allocation:** Distribution of network and compute resources to meet demand

- **Security Orchestration, Automation, and Response (SOAR):** Tools and processes that automate security operations

- **Traffic Classification:** Categorizing network traffic based on application or behavior

- **Virtualization:** Creating virtual versions of physical resources

- **Zero-Day Attack:** Cyberattack exploiting unknown vulnerabilities

Closing Remarks

This chapter equips readers with a road map for continued learning and a clear understanding of essential terminology in AI-driven SDN and networking. By exploring these resources and mastering the glossary terms, professionals and enthusiasts can deepen their expertise and confidently navigate the evolving landscape of intelligent network management.

Index

GPSR Compliance
The European Union's (EU) General Product Safety Regulation (GPSR) is a set
of rules that requires consumer products to be safe and our obligations to
ensure this.

If you have any concerns about our products, you can contact us on

ProductSafety@springernature.com

In case Publisher is established outside the EU, the EU authorized
representative is:

Springer Nature Customer Service Center GmbH
Europaplatz 3
69115 Heidelberg, Germany